Thermal and Mineral Waters

Thermal and Mineral Waters

Editor

Manoj Kumar

Thermal and Mineral Waters

Edited by **Manoj Kumar**

Printed in 2017

ISBN: 978-1-68117-157-9

Library of Congress Control Number: 2015936570

© 2016 by
SCITUS Academics LLC,
616, Corporate Way, Suite 2, 4766,
Valley Cottage, NY 10989

www.scitusacademics.com

Contents

Preface

Mineral water is water from a mineral spring that contains various minerals, such as salts and sulfur compounds. Mineral water may be effervescent (i.e., "sparkling") due to contained gases. Traditionally, mineral waters were used or consumed at their spring sources, often referred to as "taking the waters" or "taking the cure," at places such as spas, baths, or wells. The term spa was used for a place where the water was consumed and bathed in; bath where the water was used primarily for bathing, therapeutics, or recreation; and well where the water was to be consumed. Thermal analysis plays a specific role in the identification and quantitative determination of mineral components of rocks. In spite of the fact that minerals were the first group of materials studied regularly by using thermoanalytical methods, the potential offered by these methods is still not fully utilized in the field of earth sciences. The range of thermoanalytical methods applied in earth sciences is rather wide. Most works are based on DTA. DTA data provide indirect analytical information on a material and the quantification of a reaction is limited.

Editor

Thermal Characteristics and Bacterial Diversity of Forest Soil in the Haean Basin of Korea

Heejung Kim[1], Jin-Yong Lee[2], and Kang-Kun Lee[1]

[1]School of Earth and Environmental Sciences, Seoul National University, Seoul 151-747, Republic of Korea

[2]Department of Geology, Kangwon National University, Chuncheon 200-701, Republic of Korea

ABSTRACT

To predict biotic responses to disturbances in forest environments, it is important to examine both the thermophysical properties of forest soils and the diversity of microorganisms that these soils contain. To predict the effects of climate change on forests, in particular, it is essential to understand the interactions between the soil surface, the air, and the biological diversity in the soil. In this study, the temperature and thermal properties of forest soil at three depths at a site in the Haean basin of Korea were measured over a period of four months. Metagenomic analyses were also carried out to ascertain the diversity of microorganisms inhabiting the soil. The thermal diffusivity of the soil at the study site was $5.9 \times 10^{-8}\,\mathrm{m^2 \cdot s^{-1}}$. The heat flow through the soil resulted from the cooling and heating processes acting on the surface layers of the soils. The heat productivity in the soil varied through time. The phylum Proteobacteria predominated at all three soil depths, with members of Proteobacteria forming a substantial fraction (25.64 to 39.29%). The diversity and richness of microorganisms in the soil were both highest at the deepest depth, 90 cm, where the soil temperature fluctuation was the minimum.

INTRODUCTION

The health of forests has been a key concern in recent years because of natural disasters caused by environmental pollution [1–4], climate change [5–8], and ecological destruction [9–11]. Furthermore, deforestation and the degradation of natural forest environments have been central topics in United Nations venues such as the Framework Convention on Climate Change (UNFCCC), the Convention on Biological Diversity (UNCBD), and the Convention to Combat Desertification (UNCCD), marking the integrity of forest ecosystems as one of the most pressing ecological issues worldwide.

Forests provide a wide variety of important ecosystem services, including headwater conservation [12], water purification [13], soil erosion prevention [14, 15], landslide prevention [16, 17], carbon dioxide absorption [18], atmospheric purification [19, 20], conservation of native animal species diversity [21], and the protection of woodland animals [22].

Forest area of the Republic of Korea (hereafter Korea) was 6,369,000 ha in 2010, accounting for 63.7% of its total land area—the fourth-highest proportion among OECD countries. However, its forest area has been shrinking, falling from 6,640,839 ha in 1974 to 6,394,000 ha in 2005, 6,382,000 ha in 2007, and 6,370,000 ha in 2009 [23]. This decrease has been caused mainly by increasing demand for land, itself a result of increasing population and numerous development projects [24]. Even as Korea's forest area continues to decrease,

Thermal Characteristics and Bacterial Diversity of Forest Soil in the...

3

however, the forest area of other advanced countries is increasing. In Germany, a model country for forest preservation, the forest area has increased from 10,740,000 ha to 11,076,000 ha [25]. In the United States, it has increased from 225,993,000 ha to 303,089,000 ha, an increase of 77,096,000 ha [25, 26].

In this study, we investigated the thermal characteristics of forest soil at a site in the Haean basin of Korea. Forests are susceptible to climate change, and the evidence for global climate change has been reported in many studies [27–31]. We also analyzed the heat transfer characteristics of the soil, using time-series temperature data from the winter when the surface layer of soil in the forest freezes. Finally, we assessed the microbial diversity in the soil, at three sampling depths, using metagenomic analyses.

Heat transfer in forest soil is very complex and depends on factors such as conduction, heat production caused by phase transitions undergone by the water in the soil, and transfer of water and vapor between the soil and the atmosphere. Heat conduction is a key mechanism causing heat transfer within soil and is particularly important during the winter, although nonconductive heat transfer driven by the convection of groundwater can also occur if there is a suitable temperature gradient in the groundwater. The thermal diffusivity of the soil at the study site was $5.9 \times 10^{-8} \, \text{m}^2\text{s}^{-1}$. The heat flow through the soil resulted from the cooling and heating processes acting on the surface layers of the soils. The phylum Proteobacteria dominated at all three soil depths, with members of Proteobacteria forming a substantial fraction (25.64 to 39.29%) of the 16S rRNA gene sequences in all samples. In this study, in order to determine the thermal characteristics of forest soil, we used soil temperature measurements to investigate the heat transfer process by analyzing the temperature, thermal diffusivity, and heat production rate of the soil.

METHODS AND MATERIALS

Study Site

The Haean basin is located in the central region of the Korean peninsula, spanning latitudes of 38°15′–38°20′N and longitudes of 128°15′–

128°10′E (Figure 1). It is 19 km northwest of Inje-gun, Gangwon-do, and 26 km northeast of Yanggu-eup, Yanggu-gun, and is part of the Haean-myeon, Yanggu-gun, and Gangwon-do administrative districts, which border the basin. This is a rugged, mountainous region, but the Haean basin itself is a relatively flat, oval-shaped area.

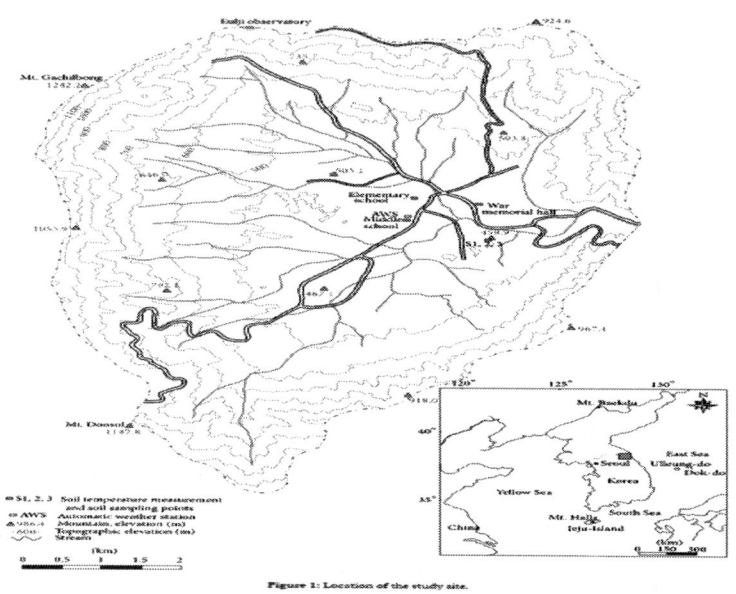

Figure 1: Location of the study site.

Figure 1: Location of the study site.

   The basin is located in the northeast of the Gyeonggi metamorphic rock complex, a region composed of Precambrian metamorphic rock (gneiss, mica-schist, and quartzite) and Jurassic igneous rock (granite) that is interpenetrated with metamorphic rock [32, 33] (Figure 2). The basin was formed by differential erosion [32, 34]; the outskirts of the basin are predominantly the harder metamorphic rocks, whereas the interior of the basin is predominantly granite, which is more vulnerable to weathering. Much of the granite is thus exposed as saprolite, which gradually turns into granodiorite and diorite [32, 35], and the soil in the basin is derived from these rocks. The basin varies in altitude from 400 m to 1,304 m, with an average altitude difference of 400–500 m between the basin bed and surrounding ridge [36].

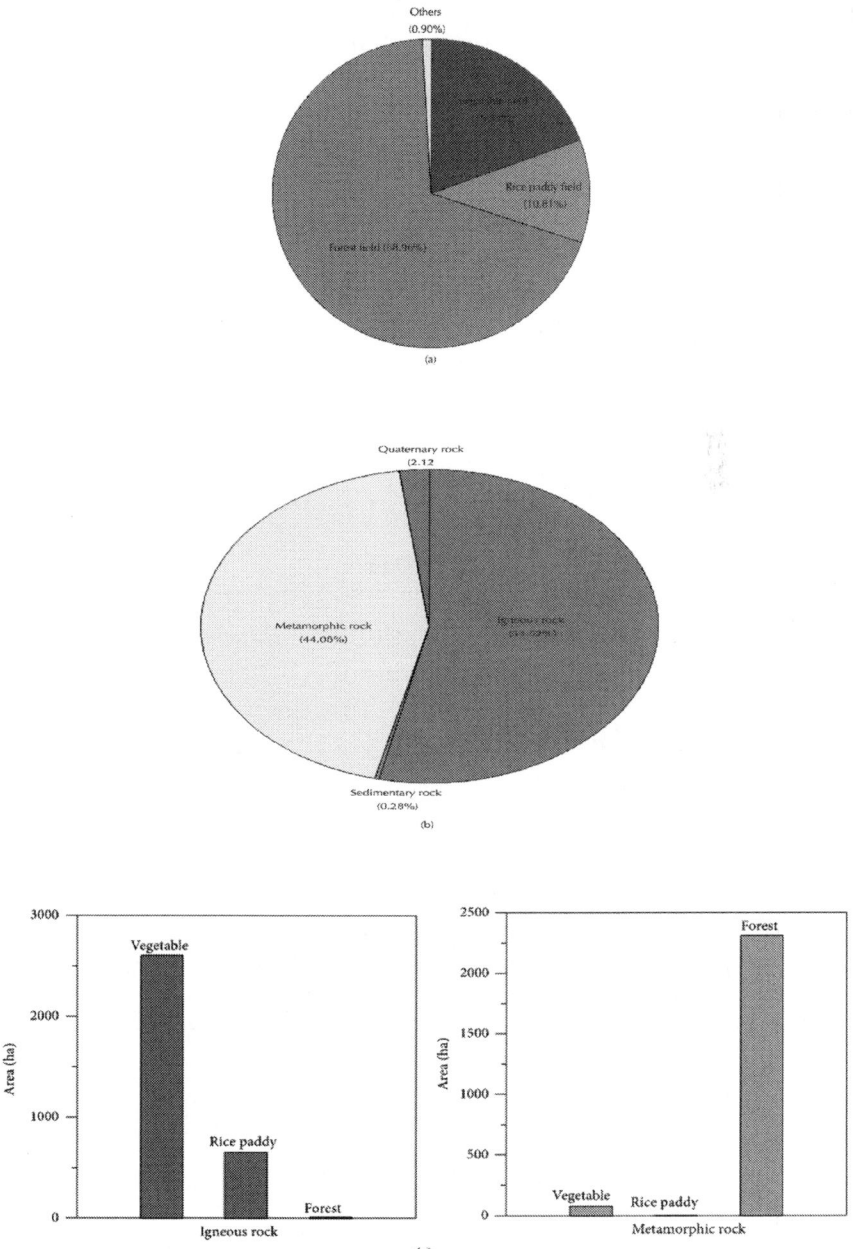

Figure 2: Land uses and soil parent materials in the Haean basin, Korea.

The average slope for the basin area as a whole is 11°, but the steepest slopes are generally in the vicinity of the surrounding ridge, with an average slope of 20°, whereas the basin bed has an average slope of only 5°. The slopes of the basin decline monotonically from ridge to bed, and thus, the basin is bowl-shaped and concave. Land use in the basin is varied; the higher altitudes, with steeper slopes, are forested, whereas the flatter areas at middle and low altitudes consist of fields and rice paddies (Figure 2). Overall, 70% of the Haean basin consists of forest. However, farming area in the basin is increasing, and so the forest area is decreasing [33, 36]. From 2002 to 2011, the maximum and minimum of annual mean air temperatures of the basin were 25.3°C and −11.5°C, respectively, and their mean was 10.1°C [36].

Soil Temperature, Thermal Diffusivity, Heat Flow, and Heat Production

Soil temperature was measured every 30 minutes using an iButton (Maxim, model DS1921G) that was installed at depths of 30, 60, and 90 cm in the Haean basin forest soil. This device is able to measure temperatures from −40°C to 80°C with a resolution of 0.01°C. Measurements were taken without interruption from November 11, 2011, to February 14, 2012, which is winter season in Korea. The temperature ranges from −1.6 to 8.4°C at 30 cm depth, −1.1 to 7.9°C at 60 cm depth, and 0.8 to 9.6°C at 90 cm depth, respectively.

From these measurements, other thermophysical properties of the soil were derived using standard methods summarized here. The law of conservation of energy can be expressed as

$$\frac{\partial}{\partial t}[cT] + \nabla \cdot \vec{q} = A,$$

(1)

where c = heat capacity, T = temperature (°C), $\vec{q}$ = heat flux, and A = heat production rate.

Heat flow, according to Fourier's law, is expressed by

$$\vec{q} = -k\nabla T.$$

(2)

The heat transfer equation can then be derived from (1) and (2) to yield

$$\nabla^2 T + \frac{A}{k} = \frac{1}{\alpha}\frac{\partial T}{\partial t},$$

(3)

where k = thermal conductivity and α = thermal diffusivity.

When (3) is solved, with an initial temperature of 0°C, for the surface temperature T over the time interval (0, t) in a semi-infinite medium in which the production and extinction of soil heat do not occur, its solution has been found to be the convolution of the surface temperature function and the heat transfer function (f_τ) [37]:

$$T(z,t) = \int_0^t T(0,\tau)\, f_T(t-\tau,z)\, d\tau.$$

(4)

Here, the heat transfer function is

$$f_T(t,z) = \frac{z}{2\,[\pi\alpha t^3]^{1/2}}\exp\left(-\frac{z^2}{4\alpha t}\right).$$

(5)

Accordingly, the change in temperature of subsurface soil, which occurs only by heat conduction in a uniform medium without production or extinction of soil heat, can be calculated using (4) and (5). In addition, since we performed soil temperature measurements at discrete depths, a finite-difference method approximation to (2) can be used to calculate heat flow by depth as

$$q = -k\frac{T(z_{i+1},t) - T(z_i,t)}{z_{i+1} - z_i}.$$

(6)

Here, the heat production rate at $[z_i, z_{i+1}] \times [t^j, t^{j+1}]$, a unit section in depth and time, can be obtained using Roth and Boike's [38] equation [39]. Consider

$$A = \frac{1}{[t^{j+1} - t^j][z_{i+1} - z_i]}$$
$$\times \int_{t^j}^{t^{j+1}} \int_{z_i}^{z_{i+1}} A(z,t)\,dz\,dt = \frac{1}{[t^{j+1} - t^j][z_{i+1} - z_i]}$$
$$\times c \left[\int_{z_i}^{z_{i+1}} \{T(z,t^{j+1}) - T(z,t^j)\}\,dz \right.$$
$$\left. - \alpha \int_{t^j}^{t^{j+1}} \{T'(z_{i+1},t) - T'(z_i,t)\}\,dt \right],$$

(7)

Where T′ = differential difference according to the depth of soil heat.

Metagenomic Analysis

Soil samples were collected for microbial analysis from three depths (30, 60, and 90 cm) from the study site in December, 2012. DNA was extracted from the samples using the FastDNA SPIN Kit for Soil (MP Biomedicals, #116560-200). 16S rRNA genes were amplified by ChunLab (Seoul, Korea) using PCR with forward and reverse primers prior to sequencing [40, 41]. The amplification conditions for PCR were (i) an initial denaturation step at 94°C for 5 min, (ii) 30 cycles of denaturation at 94°C for 30 sec followed by annealing at 55°C for 45 sec, and (iii) an extension step at 72°C for 90 sec.

Pyrosequencing of the amplified 16S rRNA was then conducted using a 454 GS FLX Titanium Junior (Roche, NJ, USA) by ChunLab (Seoul, Korea). Distinct sequences were deposited in the Sequence Read Archive (NCBI) by ChunLab. Bacterial community structures were analyzed using operational taxonomic units (OTUs). Distance matrices were used to define OTUs for calculation of the abundance-based coverage estimator (ACE), the Chao 1 richness estimator [42], Shannon and Simpson diversity indices, and rarefaction curves. Each sequence was identified by comparing it with sequences in the EzTaxon-extended database (ChunLab, eztaxon-e.org) using BLASTN searches and pairwise similarity comparisons [43].

Phylogenetic Analysis

For phylogenetic analysis of the dominant bacterial communities present in the forest soil samples, we used the 15 OTUs that were present at an abundance of at least 1%, out of the 274 OTUs (5,544 sequences) that appeared at all three sampling depths. Additionally, we used 34 OTUs that were present at an abundance of at least 0.5% in the individual samples from depths of 30 cm (10 OTUs), 60 cm (15 OTUs), and 90 cm (9 OTUs). Sequences of these 49 dominant OTUs and their related neighbors were downloaded, together with those ofDesulfurococcus kamchatkensis 1221n (NR_074374), Acidilobus aceticus 1904 (NR_041774), and Halobaculum gomorrense (L37444), which were used as outgroups.

We aligned the 49 dominant sequences with reference sequences using BioEdit version 7.0.9.0 [44]. Phylogenetic trees were constructed from the aligned sequences using three methods: (1) neighbor-joining (NJ [45]) using the Kimura 2-parameter distance model, (2) maximum likelihood (ML) using the Tamura-Nei distance model, and (3) Bayesian inference (BI) by the Markov Chain Monte Carlo (MCMC) method using the software package Molecular Evolutionary Genetics Analysis (MEGA) version 5.10 [46] and MrBayes version 3.1.2 [47]. The stability of branches was assessed using a bootstrap analysis with 1,000 replicates.

RESULTS AND DISCUSSION

Soil Temperature and Thermal Diffusivity

The mean soil temperatures, with associated standard deviations, at the three soil depths measured are shown in Table 1, while changes in the soil temperature through time are shown in Figure 3. In general, shallower soil depths are expected to show greater temperature variation. At our study site, however, the standard deviation of the temperate measurements was greatest at a depth of 60 cm. The temperature variation of shallow depth soils is generally determined by the diurnal air temperature fluctuations, wind speed, and the soil constituents, which include organic matter, mineral, water, and air. In comparison with soil with much organic matter, mineral soil tends to have higher thermal conductivity, thermal admittance, and thermal diffusivity [48, 49]. In this site, more organic matters were found at 30 cm depth than at 60 cm depth. As a consequence, a mineral soil at

60 cm depth tends to undergo large fluctuations of temperature than 30 cm depth.

Table 1: Summary statistics for temperature measurements from the study site. S1, S2, and S3 indicate the three soil depths at which measurements were taken

Point	Depth (cm)	Mean temperature (°C)	Standard deviation (°C)
S1	30	1.30	2.63
S2	60	1.99	2.65
S3	90	4.12	2.59

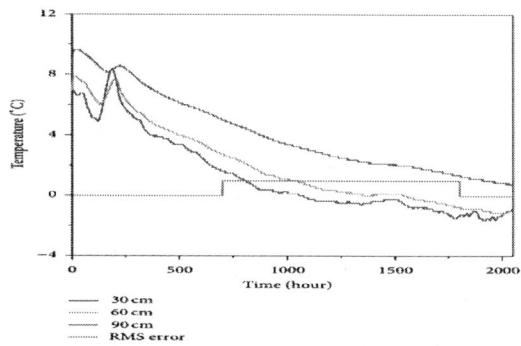

Figure 3: Soil temperatures at the study site for depths of 30, 60, and 90 cm from November 11, 2011, to February 14, 2012.

The volumetric soil heat capacity can be determined by sum of the specific heats of the soil constituents [50]. The differences of soil constituents make the variation of the soil temperature through the depths. Furthermore, the shallower soils are covered to vary with vegetation. The above assessment of soil temperature variation, while not intended to be comprehensive, suggests that temperature of shallower soil might be affected by the above factors more than 60 cm depth. The increase in temperature variation in 30 cm depth was regarded as a primary result of the organic dry soil rapidly cooling and heating by air temperature fluctuations and wind speed. Moreover, the vegetations might be able to extract moisture from 30 cm depth so it can show a temperature variation.

Changes in soil temperature can essentially be divided into four periods of time: the isothermal, cold, warming, and thawing periods [38]. In the isothermal period, the soil temperature is almost constant because the soil is in the process of freezing. In the cold period, the soil maintains a very low temperature after having completely frozen. In the warming period, the soil temperature increases prior to the initiation of melting. Finally, the thawing period spans from the initiation of melting, which starts at the ground surface, to the completion of melting. The data in this study were collected during the isothermal period and the subsequent cold period. Accordingly, they show a decreasing soil temperature trend that is accounted for by seasonality at the study site.

Since soil temperatures in the cold period are below freezing by definition, phase transitions of soil water are unimportant, resulting in a heat transfer process that mainly depends on heat conduction [51]. Therefore, subsurface temperatures can be estimated from the ground surface temperature using (4) and (5) and the thermal diffusivity can then be determined, allowing the differences between estimated and actual temperatures to be minimized. There was no existing ground surface temperature data for our study site; therefore, instead we estimated the soil temperature at 60 cm depth using the measured soil temperature at 30 cm depth and then calculated the RMS error between the estimated and measured 60 cm depth temperatures to determine the thermal diffusivity (Figure 4). This calculation was based on temperature measurements taken during the cold period, when the soil temperature was below 0°C (marked by the red line in Figure 3). The RMS error analysis revealed that the optimal thermal diffusivity was $5.9 \times 10^{-8} \, m^2 s^{-1}$ at a depth of 60 cm. The thermal diffusivity of a previous study was $4.0 \times 10^{-7} \, m^2 s^{-1}$ in Mt. Jumbong forest (10 cm depth soil [52]), near our study site. This value is greater than ours, which indicates the thermal diffusivity of forest soil is affected by not only soil constituents but also soil depth.

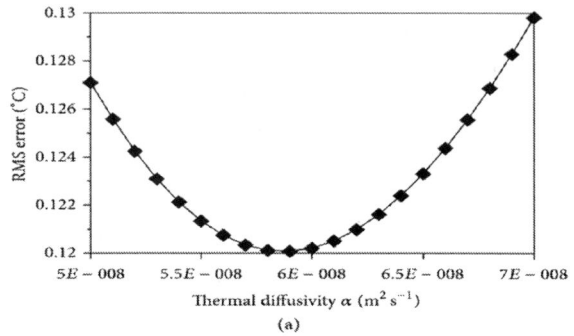

(a)

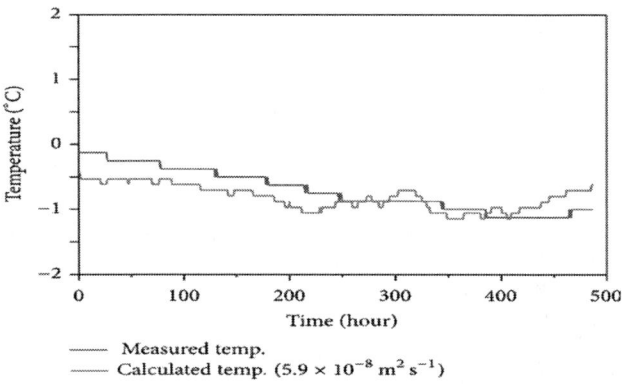

—— Measured temp.
—— Calculated temp. $(5.9 \times 10^{-8} \, m^2 \, s^{-1})$

Figure 4: RMS errors between the measured and predicted temperatures to the thermal diffusivity values for (a) a soil depth of 60 cm. Panel (b) shows best fits to this data. The RMS errors were calculated using measurements taken during the period when conduction was the dominant mode of heat transfer.

Heat Flow and Latent Heat

Heat flow at the ground surface can be calculated using a finite-difference method approximation as in (6). The heat flow through time, obtained by applying the previously determined thermal diffusivity of $5.9 \times 10^{-8} \, m^2 s^{-1}$ in this equation, is depicted in Figure 4 with the soil's average heat flow of $2.2 \times 10^6 \, Jm^{-3}K^{-1}$. In general, negative heat flow means that the local soil temperature is decreasing, while positive heat flow means it is increasing [53]. While heat transfer by heat conduction plays an important role in the process of transferring heat into the subsurface, in forest soil heat transfer by the phase change and convection of pore water also plays a central role. The production and extinction of latent heat occur during phase changes between ice, water, and vapor. In addition, water and vapor can move through the soil by convection and diffusion and can then cool and refreeze, thereby discharging latent heat in a new location [54]. Therefore, the effects of this process of phase change, convection, and diffusivity can be quantified as a heat production rate, which can be estimated using (7).

The integral in (7) can be calculated by numerical differentiation by the depth of temperature and then numerical integration by the time and depth. Since numerical differentiation uses centered finite differences, soil temperature data at four depths $[Z_{i-1}, Z_i, Z_{i+1}, Z_{i+2}]$ is needed in order to calculate the heat production rate at a single section $[Z_i, Z_{i+1}]$. We measured temperatures at depths of 30, 60, and 90 cm, and therefore the heat production rate at a depth of 60 cm could be calculated. The calculated change in heat production rate over time is shown in Figure 5. Since the duration of the monitoring period was relatively short, heat flow within the soil appeared essentially constant. Accordingly, the changes in the heat flow we observed are presumably due to seasonal climatic variation and the local cooling and heating processes at the ground surface.

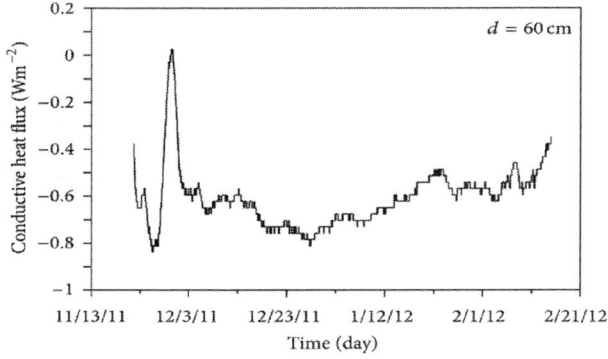

Figure 5: Heat flux estimated by the finite-difference approximation for 60 cm depth.

Bacteria Richness and Diversity Indices

To determine rarefaction curves, richness, and diversity, we identified operational taxonomic units (OTUs) in each sequencing read. The rarefaction analysis of bacterial communities derived from the three different sampling depths (S1, S2, and S3) is depicted in Figure 6. In addition, a comparison of the rarefaction analysis with the number of OTUs estimated by the Chao 1 richness estimator revealed that 73.6 to 78.6% of the estimated taxonomic richness was covered by the sequencing effort (Table 2). The Shannon index of diversity (H′) was also determined for all samples (Table 2) and ranged from 6.40 to 6.82 among the different depths. Bacterial diversity is generally expected to decrease with increasing soil depth. Here, however, the Shannon diversity index H′ was highest at the deepest depth (S3). The shallower soil tends to undergo larger and more rapid temperature responses to surface temperature fluctuations. The temperature fluctuation is often an important limitation of the soil biological activity. Ballard [55] found that changes in temperature are likely to be significant in biological activity and its little change may enhance the biological diversity and richness. The higher bacterial diversity at that depth (90 cm) is consistent with the least temperature variation and thus it can be inferred that the temperature is one of the key factors to determine the bacterial diversity and richness [56].

Table 2: Species richness estimates obtained from the soil samples

Sample	Rarefaction (number of OTUs)	Shannon index (H′)	Chao 1 (number of OTUs)	Coverage (%)
S1	1818	6.59	4331	78.6
S2	922	6.40	1625	76.4
S3	2189	6.82	6032	73.6

Thermal Characteristics and Bacterial Diversity of Forest Soil in the...

15

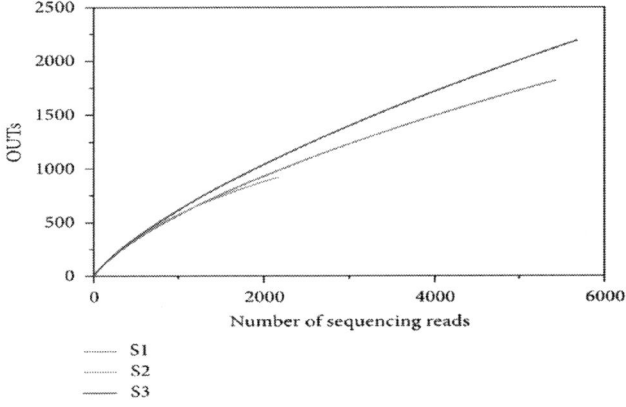

Figure 6: Rarefaction curves indicating the observed number of OTUs within the 16S rRNA gene sequences at the three depths sampled (S1 = 30 cm, S2 = 60 cm, and S3 = 90 cm).

Distribution of Taxa and Phylotypes across Samples

Figure 7 shows a phylogenetic tree generated by the neighbor-joining method that depicts the dominant bacterial relationships from the forest soil associated with the phylum Proteobacteria. Bootstrap values (>50%) based on 1,000 replicates are shown. An open circle indicates that the corresponding branch was recovered with a high bootstrap value by all three tree generation methods (neighbor-joining, maximum likelihood, and Bayesian inference), whereas a closed circle indicates that the corresponding branch was recovered only by the neighbor-joining and maximum likelihood methods. Desulfurococcus kamchatkensis 1221n (NR_074374) andAcidilobus aceticus 1904 (NR_041774) were used as the outgroups. Colors represent the dominant bacterial sequences obtained from the soil samples at various depths: 30 to 90 cm, orange; <30 cm, green; <60 cm, blue; and <90 cm, red. Figure 8 shows a phylogenetic tree generated by the neighbor-joining method that depicts the dominant bacterial relationships from the forest soil associated with phyla other than Proteobacteria.Halobaculum gomorrense (L37444) was used as an outgroup.

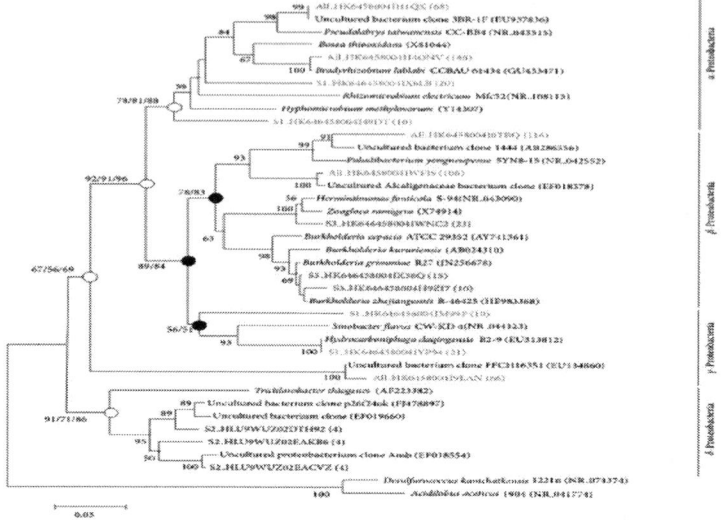

Figure 7: Neighbor-joining phylogenetic tree showing the dominant bacterial relationships associated with phylum Proteobacteria in the soil samples. The bar in the bottom represents 0.05 nucleotide substitutions per nucleotide position.

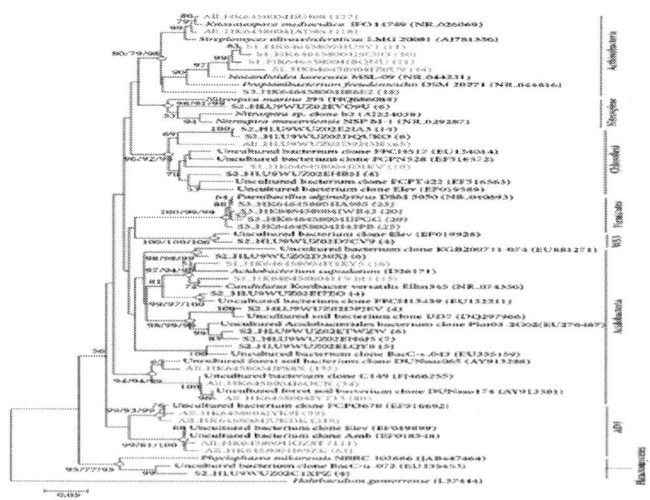

Figure 8: Neighbor-joining phylogenetic tree showing the dominant bacterial relationships associated with phyla other than Proteobacteria in the soil

samples. The bar in the bottom represents 0.05 nucleotide substitutions per nucleotide position.

The 5,544 sequences were affiliated with 40 phyla across the entire data set. The dominant phyla within S1 were Proteobacteria, Acidobacteria, Actinobacteria, Chloroflexi, Gemmatimonadetes, Planctomycetes, Nitrospirea, Cyanobacteria, Firmicutes, and Bacteroidetes, representing 39.29, 27.03, 7.65, 3.94, 2.04, 1.55, 1.27, 1.18, 1.05, and 0.96% of the sequences, respectively. The dominant phyla within S2 were Acidobacteria, Proteobacteria, Chloroflexi, Actinobacteria, Gemmatimonadetes, Nitrospirea, Verrucomicrobia, Planctomycetes, Thermobaculum_p, and Bacteroidetes, representing 32.55, 25.64, 7.37, 6.63, 4.56, 3.31, 2.39, 1.70, 1.34, and 1.01% of the sequences, respectively. The dominant phyla across S3 were Proteobacteria, Acidobacteria, Actinobacteria, Firmicutes, Chloroflexi, Gemmatimonadetes, Nitrospirea, Verrucomicrobia, Cyanobacteria, Planctomycetes, and Bacteroidetes, representing 31.91, 21.21, 14.35, 4.86, 3.23, 2.13, 1.74, 1.55, 1.32, 1.28, and 1.12% of the sequences, respectively.

The phylum Proteobacteria was thus predominant across the S1, S2, and S3 samples taken together, and its members comprised a substantial percentage (25.64 to 39.29%) of the 16S rRNA gene sequences at all of the depths sampled. The phylum Proteobacteria is divided into 4 subgroups (α, β, γ and δ-Proteobacteria). Members of the phylum Acidobacteria were also very common in all samples, comprising a substantial percentage (21.21 to 32.55%) of the 16S rRNA gene sequences at all of the depths sampled. This abundance of Acidobacteria is in accord with other studies [57] of the composition of soil-derived bacterial communities from a variety of environments, such as forests, grasslands, and agricultural areas. Although we covered 5,544 sequences, we did not examine the full extent of bacteria richness at the phylum level within various depth soils. In this case, the S2 soil showed a lower estimated bacterial diversity than the S1 and S3 soils. Thus, it can be inferred that the S1 soil is more plentiful of organic matter than S2 and S3 soil is more stable than S1 and S2 soils with respect to temperature variation.

CONCLUSIONS

Thermal characteristics and microbial diversity were examined in the forest soil of the Haean basin of Korea. The thermal diffusivity of the soil was $5.9 \times 10^{-8} \, m^2 s^{-1}$ at a depth of 60 cm. Heat flow was negative at depths where the soil temperature was decreasing, whereas heat increased when the internal temperature of the soil was higher than the ground surface temperature. Analysis of microbial diversity revealed that diversity and species richness varied with depth. In particular, the microbial species richness was highest at the deepest depth sampled, 90 cm, a result that is contrary to the usual relationship between microbial species richness and soil depth. The reason for this appeared to be that the soil at 90 cm depth maintained a relatively steady temperature, allowing a high species richness and diversity, compared to the shallower depths (30~60 cm) where the temperature fluctuated sharply. Studies of the thermophysical properties and microbial diversity of forest soils are valuable since they can help predict the biological diversity of forest ecosystems, allowing us to evaluate the effects of climate change on these ecosystems in the coming years.

ACKNOWLEDGMENTS

This research was supported by the Basic Science Research Program through the National Research Foundation of Korea (NRF), the Ministry of Education, Science and Technology (NRF-2011-0007232), and the Advanced Technology for Groundwater Development Program (code 11 Technology Innovation C05) of the MOLIT and the KAIA of Korea.

REFERENCES

1.   W. H. Smith, Interactions between Air Contaminants and Forest Ecosystems, Springer, 1981.
2.   E.-D. Schulze, "Air pollution and forest decline in a spruce (Picea abies) forest," Science, vol. 244, no. 4906, pp. 776–783, 1989.
3.   G. E. Taylor Jnr, D. W. Johnson, and C. P. Andersen, "Air pollution and forest ecosystems: a regional to global perspective," Ecological Applications, vol. 4, no. 4, pp. 662–689, 1994.

4.   K. P. Beckett, P. H. Freer-Smith, and G. Taylor, "Urban woodlands: their role in reducing the effects of particulate pollution," Environmental Pollution, vol. 99, no. 3, pp. 347–360, 1998.

5.   J. Pastor and W. M. Post, "Response of northern forests to CO_2-induced climate change," Nature, vol. 334, no. 6177, pp. 55–58, 1988.

6.   J. T. Overpeck, D. Rind, and R. Goldberg, "Climate-induced changes in forest disturbance and vegetation," Nature, vol. 343, no. 6253, pp. 51–53, 1990.

7.   V. H. Dale, L. A. Joyce, S. McNulty et al., "Climate change and forest disturbances," BioScience, vol. 51, no. 9, pp. 723–734, 2001.

8.   Y. Malhi, J. T. Roberts, R. A. Betts, T. J. Killeen, W. Li, and C. A. Nobre, "Climate change, deforestation, and the fate of the Amazon," Science, vol. 319, no. 5860, pp. 169–172, 2008.

9.   B. Ulrich, "A concept of forest ecosystem stability and of acid deposition as driving force for destabilization," in Accumulation of Air Pollutants in Forest Ecosystems , Proceedings of a Workshop, B. Ulrich and J. Pankrath, Eds., pp. 1–9, Springer, Gottingen, Germany, 1983.

10.   F. H. Bormann, "Air pollution and forests: an ecosystem perspective," BioScience, vol. 35, no. 7, pp. 434–441, 1985.

11.   T. Magura, "Carabids and forest edge: Spatial pattern and edge effect," Forest Ecology and Management, vol. 157, no. 1-3, pp. 23–37, 2002.

12.   J. Krishnaswamy, M. Bonell, B. Venkatesh et al., "The groundwater recharge response and hydrologic services of tropical humid forest ecosystems to use and reforestation: support for the infiltration-evapotranspiration trade-off hypothesis," Journal of Hydrology, vol. 498, pp. 191–209, 2013.

13.   K. N. Ninan and M. Inoue, "Valuing forest ecosystem services: what we know and what we don't,"Ecological Economics, vol. 93, pp. 137–149, 2013.

14.   J. H. Patric, "Soil erosion in the Eastern forest," Journal of Forestry, vol. 129, no. 10, pp. 671–677, 1976.

15.   P. Borrelli and B. Schütt, "Assessment of soil erosion sensitivity and post-timber-harvesting erosion response in a mountain

environment of Central Italy," Geomorphology, vol. 204, pp. 412–424, 2014.

16.   C. B. Brown and M. S. Sheu, "Effects of deforestation on slopes," Journal Geotechnical Engineering Division, vol. 26, no. 101, pp. 147–165, 1975.

17.   A. J. C. Collison and M. G. Anderson, "Using a combined slope hydrology/stability model to identify suitable conditions for landslide prevention by vegetation in the humid tropics," Earth Surface Processes and Landforms, vol. 21, no. 8, pp. 737–747, 1996.

18.   G. McPherson, J. R. Simpson, P. J. Peper, S. E. Maco, and Q. Xiao, "Municipal forest benefits and costs in five US cities," Journal of Forestry, vol. 103, no. 8, pp. 411–416, 2005.

19.   E. G. McPherson, D. Nowak, G. Heisler et al., "Quantifying urban forest structure, function, and value: the Chicago Urban Forest Climate Project," Urban Ecosystems, vol. 1, no. 1, pp. 49–61, 1997.

20.   K. I. Scott, E. G. McPherson, and J. R. Simpson, "Air pollutant uptake by Sacramento's urban forest,"Journal of Arboriculture, vol. 24, no. 4, pp. 224–231, 1998.

21.   E. Führer, "Forest functions, ecosystem stability and management," Forest Ecology and Management, vol. 132, no. 1, pp. 29–38, 2000.

22.   Å. Berg, B. Ehnstrom, L. Gustafsson, T. Hallingback, M. Jonsell, and J. Weslien, "Threatened plant, animal, and fungus species in Swedish forests: distribution and habitat associations," Conservation Biology, vol. 8, no. 3, pp. 718–731, 1994.

23.   Korea Forest Service, Basics statistics of forests, http://www.forest.go.kr.

24.   KOSIS, "Statistics Korea," 2014, http://kostat.go.kr/.

25.   FAO (Food and Agriculture Organization of the United Nations), "FAOSTAT forestry database,"http://www.fao.org/home/en/.

26.   US Forest Service, "United States of Forest Service," http://www.fs.fed.us/.

27.   T. Karjalainen, A. Pussinen, J. Liski, et al., "An approach towards an estimate of the impact of forest management and climate change on the European forest sector carbon budget: Germany

as a case study," Forest Ecology and Management, vol. 162, no. 1, pp. 87–103, 2002.

28. J. P. Siry, F. W. Cubbage, and M. R. Ahmed, "Sustainable forest management: global trends and opportunities," Forest Policy and Economics, vol. 7, no. 4, pp. 551–561, 2005.

29. C. B. Schmitt, N. D. Burgess, L. Coad et al., "Global analysis of the protection status of the world›s forests," Biological Conservation, vol. 142, no. 10, pp. 2122–2130, 2009.

30. H. Eva, S. Carboni, F. Achard et al., "Monitoring forest areas from continental to territorial levels using a sample of medium spatial resolution satellite imagery," ISPRS Journal of Photogrammetry and Remote Sensing, vol. 65, no. 2, pp. 191–197, 2010.

31. P. Meyfroidt and E. F. Lambin, "Global forest transition: prospects for an end to deforestation," Annual Review of Environment and Resources, vol. 36, pp. 343–371, 2011.

32. Y. S. Kwon, H. H. Lee, U. Han, and W. H. Kim, "Terrain analysis of Haean basin in terms of earth science," Journal of Korea Earth Science Society, vol. 11, pp. 236–241, 1990 (Korean).

33. National Academy of Agricultural Science, "Korean Soil Atlas," http://soil.rda.go.kr.

34. J. Y. Lee, "Importance of hydrogeological and hydrologic studies for Haean basin in Yanggu," Journal of the Geological Society of Korea, vol. 45, pp. 405–414, 2009 (Korean).

35. H. Kim, K. K. Lee, and J. Y. Lee, "Numerical verification of hyporheic zone depth estimation using streambed temperature," Journal of Hydrology, vol. 511, pp. 861–869, 2014.

36. H. Kim, J. Y. Lee, and K. K. Lee, "Spatial and temporal variations of groundwater—stream water interaction in an agricultural area, case study: Haean basin, Korea," Research Journal of Earth and Planetary Sciences, vol. 2, no. 2, pp. 71–82, 2013.

37. H. S. Carslaw and J. C. Jaeger, Conduction of Heat in Solids, Oxford University Press, Oxford, UK, 2nd edition, 1986.

38. K. Roth and J. Boike, "Quantifying the thermal dynamics of a permafrost site near Ny-Ålesund, Svalbard," Water Resources Research, vol. 37, no. 12, pp. 2901–2914, 2001.

39. U. Han, C. K. Lee, S. Jeong, B. Y. Lee, and S. H. Nam, "The studies of the temperature and thermal properties of the active layer at

the Seang Station, Antarctica," Journal of the Geological Society of Korea, vol. 42, no. 4, pp. 577–586, 2006 (Korean).

40. T. Unno, J. Jang, D. Han, et al., "Use of barcoded pyrosequencing and shared OTUs to determine sources of fecal bacteria in watersheds," Environmental Science & Technology, vol. 44, no. 20, pp. 7777–7782, 2010.

41. S. Lim, S. Kim, K.-M. Yeon, B.-I. Sang, J. Chun, and C.-H. Lee, "Correlation between microbial community structure and biofouling in a laboratory scale membrane bioreactor with synthetic wastewater," Desalination, vol. 287, pp. 209–215, 2012.

42. A. Chao and J. Bunge, "Estimating the number of species in a stochastic abundance model," Biometrics, vol. 58, no. 3, pp. 531–539, 2002.

43. J. Chun, J.-H. Lee, Y. Jung et al., "EzTaxon: a web-based tool for the identification of prokaryotes based on 16S ribosomal RNA gene sequences," International Journal of Systematic and Evolutionary Microbiology, vol. 57, no. 10, pp. 2259–2261, 2007.

44. T. A. Hall, "BioEdit: a user-friendly biological sequence alignment editor and analysis program for Windows 95/98/NT," Nucleic Acids Symposium Series, vol. 41, pp. 95–98, 1999.

45. N. Saitou and M. Nei, "The neighbor-joining method: a new method for reconstructing phylogenetic trees," Molecular biology and evolution, vol. 4, no. 4, pp. 406–425, 1987.

46. K. Tamura, D. Peterson, N. Peterson, G. Stecher, M. Nei, and S. Kumar, "MEGA5: molecular evolutionary genetics analysis using maximum likelihood, evolutionary distance, and maximum parsimony methods," Molecular Biology and Evolution, vol. 28, no. 10, pp. 2731–2739, 2011.

47. M. Holder and P. O. Lewis, "Phylogeny estimation: traditional and Bayesian approaches," Nature Reviews Genetics, vol. 4, no. 4, pp. 275–284, 2003.

48. T. M. Ballard, T. A. Black, and K. G. McNaughton, "Summer energy balance and temperatures in a forest clearcut in southwestern British Columbia," in Proceedings of the 6th B.C. Soil Science Workshop Report, pp. 74–85, Victoria, Canada, 1977.

49. P. S. J. Verburg, W. K. P. Van Loon, and A. Lükewille, "The CLIMEX soil-heating experiment: soil response after 2 years of treatment," Biology and Fertility of Soils, vol. 28, no. 3, pp. 271–276, 1999.

50. C. L. Mayocchi and K. L. Bristow, "Soil surface heat flux: some general questions and comments on measurements," Agricultural and Forest Meteorology, vol. 75, no. 1–3, pp. 43–50, 1995.

51. J. Boike, K. Roth, and P. P. Overduin, "Thermal and hydrologic dynamics of the active layer at a continuous permafrost site (Taymyr Peninsula, Siberia)," Water Resources Research, vol. 34, no. 3, pp. 355–363, 1998.

52. S. Kang, S. Kim, S. Oh, and D. Lee, "Predicting spatial and temporal patterns of soil temperature based on topography, surface cover and air temperature," Forest Ecology and Management, vol. 136, no. 1, pp. 173–184, 2000.

53. E. M. Fischer, S. I. Seneviratne, P. L. Vidale, D. Lüthi, and C. Schär, "Soil moisture-atmosphere interactions during the 2003 European summer heat wave," Journal of Climate, vol. 20, no. 20, pp. 5081–5099, 2007.

54. U. Han, C. K. Lee, S. H. Nam, B. Y. Lee, and Y. Kim, "Thermal dynamics of active layer at the Dasan station, Svalbard," Journal of the Geological Society of Korea, vol. 41, no. 1, pp. 91–100, 2005 (Korean).

55. T. M. Ballard, "Impacts of forest management on northern forest soils," Forest Ecology and Management, vol. 133, no. 1-2, pp. 37–42, 2000.

56. V. Torsvik and L. Øvreås, "Microbial diversity and function in soil: from genes to ecosystems," Current Opinion in Microbiology, vol. 5, no. 3, pp. 240–245, 2002.

57. G. Imfeld, H. Pieper, N. Shani et al., "Characterization of groundwater microbial communities, dechlorinating bacteria, and in situ biodegradation of chloroethenes along a vertical gradient," Water, Air, and Soil Pollution, vol. 221, no. 1–4, pp. 107–122, 2011.

Chapter 2

Arima Hot Spring Waters as a Deep-Seated Brine from Subducting Slab

Chiho Kusuda[1], Hikaru Iwamori[2, 3], Hitomi Nakamura[2, 3], Kohei Kazahaya[4], and Noritoshi Morikawa[4]

[1]26-1 Yamanokuchi Chitosecho-chitose, Kameoka, Kyoto 621-0002, Japan

[2]Department of Solid Earth Geochemistry, Japan Agency for Marine–Earth Science and Technology (JAMSTEC), 2-15 Natsushima-cho, Yokosuka 237-0061, Japan

[3]Department of Earth and Planetary Sciences, Tokyo Institute of Technology, 2-12-1 Ookayama, Meguro-ku, Tokyo 152-8551, Japan

[4]Geological Survey of Japan, National Institute of Advanced Industrial Science and Technology (AIST), 1-1-1 Higashi, Tsukuba 305-8567, Japan

ABSTRACT

Non-volcanic hot springs are generally believed to originate through circulation of meteoric or buried sea water heated at depth. In this study, we report the geochemical characteristics of the Arima and Takarazuka hot spring waters, known as Arima-type deep brine, in a forearc region of southwestern Japan. We examine 14 water samples to determine the levels of 12 solute elements or components and the isotopic ratios of H, He, C, O, and Sr, and we perform correlation analysis of the data to deduce the source materials and origin of the deep brine. Moreover, we perform numerical modeling of oxygen and hydrogen isotopic fractionation along subducting slabs to examine the composition of slab-derived fluid as a possible candidate of the deep brine. The results suggest that the high salinity and solute concentrations with characteristic oxygen, hydrogen, carbon, and strontium isotope compositions, as well as high $^3He/^4He$ ratios, can be explained by a dehydrated component of the subducted Philippine Sea slab. Hence, this study may provide an invaluable understanding of geofluid processes over a significant depth range.

BACKGROUND

Various geophysical and geochemical studies suggest that within the mantle wedge and crust in subduction zones, fluid fluxes may trigger seismicity and magmatism throughout the system (e.g., Iwamori 2007; Hasegawa et al. 2008). The term 'fluid' in this case includes a variety of physical substances such as aqueous fluid, supercritical fluid, and melt (or magma), which are often referred to as 'geofluids.' Studies on geofluids may provide a clue to understanding the hydrological budget in subduction zones, as well as the global water and element cycling, since subduction zones are the major injection sites of water and chemical elements existing at the surface into the Earth's interior (Jarrard 2003; Fischer 2008; Iwamori and Albarède 2008; Shinohara 2013).

With its distinct geochemical signatures, the Arima-type brine is known as a possible candidate of deep-seated geofluid. Despite its presence in non-volcanic regions in the forearc domain, in which Quaternary volcanoes do not appear (Figure 1), the oxygen and hydrogen isotope compositions have affinities to magmatic/ metamorphic fluids; thus, the presence of a deep brine distinct from near-surface water, including meteoric water and seawater, has been long argued for the origin of Arima hot springs (e.g., Matsubaya et al. 1973; Tanaka et al. 1984; Masuda et al. 1985 and the references therein; overview presented in Sakai and Matsuhisa 1996). Although a precise definition of Arima-type brine has not been presented, non-volcanic hot springs with high chlorine (approximately 40,000 ppm) and other solute contents, as well as the distinct oxygen and hydrogen isotopic ratios, occur in the Osaka to Kii areas, southwestern Japan, along large fault zones (Matsubaya 1981; Nishimura 2000), where high $^3He/^4He$ of gases occur associated with spring waters (Nagao et al. 1981; Sano and Wakita 1985; Okada et al. 1994; Matsumoto et al.2003; Umeda et al. 2006; Morikawa et al. 2008).

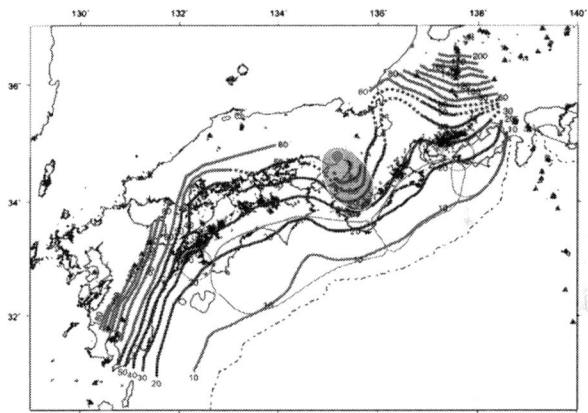

Figure 1: Locality and tectonic setting of the Arima hot springs (small red circle), modified from Hirose et al. (2008). The pink area involving Arima shows the arbitrary boundary of the 'Kinki Spot,' where anomalously high $^3He/^4He$ ratios have been observed (Sano and Wakita1985; Sano et al. 2009). Contour lines indicate the upper surface of the subducted Philippine Sea Plate, and the triangles denote Quaternary volcanoes.

On the contrary, the nature and composition of such deep geofluids have been studied through high-pressure experiments on element partitioning (e.g., Brenan et al. 1995; Keppler 1996; Kogiso et al. 1997; Kessel et al. 2005), geochemical constraints on subducting materials and arc magmas (e.g., Ishikawa and Nakamura 1992; Plank and Langmuir 1998; Pearce et al. 2005; Nakamura et al. 2008; Kimura et al. 2009), and fluid inclusions in mantle-derived xenoliths (e.g., Schiano et al.1995; Ishimaru and Arai 2008; Ionov 2010; Kawamoto et al. 2013), constraining elemental abundances and isotopic ratios of these geofluids.

In this study, we discuss the relationship between Arima-type brine and deep geofluids in a definitive manner. In particular, we examine multiple element and isotopes of the hot spring waters, including the oxygen and hydrogen isotopic composition of the slab-derived fluid, and we determine whether the characteristic O-H isotopic compositions of the Arima-type brine are explained by the slab fluid.

Multiple Elemental and Isotopic Characterization

Sample and Analytical Methods

The samples and localities are listed in Table 1. In December 2007, water and gas samples were collected from hot mineral springs and artesian wells at 14 sites in the Arima area, located on the northern slope of the Rokko Mountains where basement rocks are composed mainly of Paleozoic to Mesozoic sedimentary rocks; upper Cretaceous to Paleogene volcanic rocks, including mostly rhyolites and andesites; and Cretaceous to Paleogene granitic rocks (Arai 2007). The Arima-Takatsuki Tectonic Line and the associated faults are distributed in the studied area, along which the Arima hot spring waters flow from the surface (Kozuki 1962).

Table 1: Sample and locality list

Sample No.	Name of springs	Latitude	Longitude	Sampling date
TW-1, TG-1	Takarazuka Tibori	34° 47′51.66″	135° 22′01.32″	12 December 2007
TW-2	Takarazuka Spa	34° 48′20.22″	135° 20′13.56″	12 December 2007

AW-1, AG-1	Gosya	34° 48'12.42"	135° 12'51.60"	13 December 2007
AW-2	Tansan	34° 47'41.94"	135° 15'02.04"	13 December 2007
AW-3	Gokuraku	34° 47'46.08"	135° 14'57.12"	13 December 2007
AW-4	Uwanari	34° 47'45.96"	135° 14'57.12"	13 December 2007
AW-5, AG-5	Gosyo	34° 47'47.34"	135° 14'52.80"	13 December 2007
AW-6	Tenjin	34° 47'49.32"	135° 14'54.42"	13 December 2007
AW-7	Kosenkaku	34° 47'55.86"	135° 14'50.28"	13 December 2007
AW-8	Mint Resort Inn	34° 48'11.28"	135° 14'59.28"	14 December 2007
AW-9	Ginsuiso Ginsen south	34° 48'03.48"	135° 14'47.94"	14 December 2007
AW-10	Ginsuiso Ginsen north	34° 48'06.66"	135° 14'51.18"	14 December 2007
AW-11	Ginsuiso Kinsen	34° 48'06.42"	135° 14'50.88"	14 December 2007
AW-12	Choraku Kinsen	34° 48'12.72"	135° 14'49.62"	14 December 2007

Kusuda et al.

Kusuda et al. Earth, Planets and Space 2014 66:119, doi:10.1186/1880-5981-66-119

The studied area is within 600 m zonally and 900 m meridionally in Arima. In addition, two samples were obtained in Takarazuka, about 10 km east of Arima, where most of the springs flow naturally from shallow (40 m in depth) to deep (1500 m) reservoirs or flow paths along the faults (e.g., Kozuki 1962), with a temperature range of 19°C to 96°C (Table 1).

In order to acquire a multi-element and multi-isotope dataset for each sample, the following measurements and analyses were performed for the sampled solutions and gases: (1) Temperature, pH, and electrical conductivity (EC; Table 2) of waters were conducted *in situ* at sampling sites. (2) Concentrations of cations and anions of water samples were determined with ion chromatography (DXi-500; Thermo Scientific Dionex, CA, USA). (3) Alkalinity, in reference to the concentration of total dissolved HCO_3, was determined by titrating with H_2SO_4 up to a pH of 4.8. (4) Hydrogen and oxygen stable isotopes were measured by mass spectrometric analysis (Delta V advantage and Delta plus, Thermo-Fisher Scientific Inc., Bremen, Germany). H_2 generated by the H_2O-H_2 reduction method by 800°C metal chromium was applied for δD analyses, and CO_2 generated by the automated H_2O-CO_2 equilibration method was applied for $\delta^{18}O$ analyses. (5) The stable carbon isotopic ratio ($\delta^{13}C$) of dissolved inorganic carbon (DIC), which was conventionally represented with respect to Vienna Pee Dee Belemnite (VPDB), was measured by continuous flow isotope-ratio mass spectrometry (IRMS) with a gas chromatography system (Delta-V Advantage and GasBench II, Thermo-Fisher Scientific Inc.). The detailed procedures have been described by Takahashi et al. (2013). (6) Helium and neon concentrations and the helium isotope ratios were measured with a noble gas mass spectrometer, model MM5400 (Micromass UK Ltd., Manchester, UK). Technical details of the extraction of dissolved noble gases and the mass spectrometry, including the purification procedures, have been described by Morikawa et al. (2008). (7) Sr concentrations were measured with an inductively coupled plasma mass spectrometer (ICPMS; ELAN DRC II, Perkin Elmer, Waltham, MA, USA) after dilution of the sample with HNO_3 containing the indium internal standard. (8) Sr isotope ratios were determined with a thermal ionization mass spectrometer (TIMS; TRITON, Thermo Finnigan, Thermo-Fisher Scientific Co. Ltd., Hudson, NH, USA) after chemical separation of Sr using a 0.3-mL DCTA-pyridine cation-exchange column with a procedure modified after Birck (1986). Laboratory analyses (2) to (6) were performed at the Geological Survey of Japan, AIST; (7) and (8) were performed at the Japan Agency for Marine-Earth Science and Technology.

Table 2: Analytical results of the Arima and Takarazuka hot spring waters and gases

	TW-1 [TG-1]	TW-2	AW-1 [AG-1]	AW-2	AW-3	AW-4	AW-5 [AG-5]	AW-6	AW-7	AW-8	AW-9	AW-10	AW-11	AW-12
Temperature (°C)	22.3	25.9	32.1	19.1	95.3	88.3	80.3	88.3	61.5	33.3	19.4	19.5	29.2	32.2
pH	6.20	6.47	6.23	5.30	6.42	6.65	6.89	6.38	6.51	6.04	5.78	6.51	5.64	6.00
EC (mS/cm)	17.72	62.10	11.79	0.30	62.80	57.10	26.40	63.30	40.40	42.20	1.85	4.60	70.80	67.50
Depth (m)	800	1,500	nk	40	240	187	165	206	170	800	nk	nk	nk	nk
$\delta^{18}O_{SMOW}$ (‰)	-0.35	1.09	-7.00	-7.88	2.07	0.77	-4.09	2.22	-1.85	-3.26	-7.27	-7.08	4.98	3.61
δD_{SMOW} (‰)	-40.44	-38.51	-51.05	-50.27	-38.15	-40.50	-45.92	-37.71	-44.69	-44.31	-48.92	-48.56	-34.12	-35.11
$\delta^{13}C_{PDB}$ (%)	-1.29	-1.05	-3.83	-4.25	0.15	-2.22	-3.73	-1.05	-2.53	-5.35	-7.67	-1.82	-0.86	-13.42
F (mg/L)	0.00	1.22	1.99	0.60	0.99	0.88	1.47	1.39	1.30	1.00	0.29	0.93	0.00	0.00
Cl (mg/L)	16,666	21,439	3,392	14	25,088	20,786	8,742	25,828	14,637	15,333	435	1,294	40,033	36,602
Br (mg/L)	37.99	51.48	7.17	0.02	39.69	32.64	13.15	42.19	24.13	34.12	0.62	2.04	72.95	83.21
SO_4 (mg/L)	6.54	0.70	1.12	59.02	2.41	2.41	24.92	2.77	10.30	1.69	15.27	23.06	10.25	2.43
Li (mg/L)	38.16	40.75	6.50	0.03	35.78	30.76	14.42	36.69	22.97	20.86	0.05	2.08	51.90	33.81
Na (mg/L)	9,623	12,618	1,743	18	11,945	10,127	4,700	12,189	7,805	9,014	23	704	20,077	18,379
NH_4 (mg/L)	7.77	5.46	0.86	0.03	4.82	4.42	2.33	5.23	2.52	4.17	0.00	0.04	8.22	3.06
K (mg/L)	959	1,372	208	3	2,525	2,153	1,000	2,593	1,388	332	4	131	1,930	112
Mg (mg/L)	284	114	35	1	15	12	15	17	29	144	0	4	136	511
Ca (mg/L)	629	985	491	23	2,192	1,845	627	2,356	1,283	860	8	108	3,006	2,598
HCO_3 (mg/L)	3,803	3,154	781	30	26	203	473	137	435	1,623	93	150	1,293	214
$^3He/^4He$ (10^{-6})	7.11 [8.92]	5.46	8.98 [9.42]	10.29	3.26	3.34	3.84 [10.51]	na	6.73	5.07	7.62	na	6.69	na
$^3He/^4He$ (Ra)	5.08 [6.37]	3.90	6.41 [6.73]	7.35	2.33	2.38	2.75 [7.51]	na	4.80	3.62	5.44	na	4.78	na

⁴He (10⁻⁸)	²⁰Ne (10⁷)	⁴He/²⁰Ne	Sr (mg/L)	⁸⁷Sr/⁸⁶Sr
13.87 [11.51]	0.79 [0.58]	1.75 [19.90]	20.23	0.70926
3.90	0.43	0.91	43.34	0.70878
499.7 [646.9]	0.44 [2.83]	113.9 [228.3]	8.28	0.70854
191.10	0.18	106.15	0.12	0.70846
1.77	0.42	0.43	43.39	0.70853
2.52	0.67	0.38	37.08	na
4.08 [3.35]	0.98 [0.042]	0.42 [79.5]	11.09	na
na	na	na	41.01	na
8.29	0.89	0.93	28.88	na
17.83	1.63	1.10	48.32	0.70879
88.18	9.17	0.96	0.76	na
na	na	na	0.77	na
413.90	na	na	80.35	na
na	na	na	93.94	

nk, not known; na, not analyzed. The concentrations of noble gases are shown in [cm3 STP (noble gas)/g (sampled water)]. TG-1, AG-1, and AG-5 are the gas samples for which the He and Ne data shown in square brackets are available.

Kusuda et al.

Kusuda et al. Earth, Planets and Space 2014 66:119, doi:10.1186/1880-5981-66-119

Elemental and Isotopic Composition

The results are listed in Table 2. Both the solute concentrations and the isotopic ratios of hydrogen, helium, carbon, oxygen, and strontium exhibited wide variations. The chlorine concentration ranges from 14 to approximately 40,000 ppm (up to twice that of seawater), the lithium concentration ranges from 0.02 to 51.90 ppm (0.2 to 300 times that of seawater), and the air-normalized ^{3}He/^{4}He ratio ranges from 2.33 to 7.51 (close to that of the upper mantle represented by mid-ocean ridge basalt). In order to discuss origin of the brines that may potentially involve multiple source materials and processes, analyses based on multiple element/isotope data, as shown in Table 2, are useful for clarifying, for example, the association of elements/isotopes derived from individual sources. Such multiple element/isotope studies, including hydrocarbon species, on brine or gas from various tectonic settings such as subduction zones (e.g., Mutnovsky Volcano, Kamchatka; Zelenski et al. 2012), ocean islands (e.g., Socorro Island, Mexico; Taran et al. 2010), and continental-oceanic rifts (Salton Sea, USA; Mazzini et al. 2011) have been successfully used to identify various sources. These sources include altered basaltic oceanic crust (AOC), oceanic sediment, and continental and mantle sources, and their modification by chemical and thermal processes such as mixing, serpentinization, and near-surface cooling.

The correlation matrix, in which correlation coefficients are arranged in a matrix consisting of individual elements/isotopes to be correlated to show the correlation coefficient at their intersection cell (Figure 2), shows an overall data structure. Many of the major solute elements/ions such as Cl, Na, Li, Br, Ca, Sr, NH_4, and K have sharp positive correlations with $\delta^{18}O$ and δD, resulting in a strong correlation with the electrical conductivity (EC indicated by asterisk in Figure 2). On the contrary, ^{3}He/^{4}He and SO_4 are broadly and negatively correlated with $\delta^{18}O$, δD, and other solute elements, whereas $^{87}Sr/^{86}Sr$, $\delta^{13}C$, HCO_3, and F, as well as temperature and pH, are poorly correlated with the major solute elements. The sharp positive correlation among $\delta^{18}O$, δD, and the major solute elements is detailed below.

	$\delta^{18}O$	δD	Cl	Na	Li	EC*	Br	Ca	Sr	NH_4	K	Mg	T*	$\frac{^{87}Sr}{^{86}Sr}$	HCO_3	pH*	$\delta^{13}C$	F	SO_4	$\frac{^3He}{^4He}$
$\delta^{18}O$	1.00	0.99	0.98	0.97	0.96	0.94	0.93	0.97	0.87	0.83	0.71	0.48	0.38	0.32	0.73	0.13	0.12	-0.32	-0.58	-0.58
δD	0.99	1.00	0.97	0.97	0.95	0.91	0.94	0.89	0.88	0.83	0.65	0.53	0.30	0.36	0.26	0.08	0.08	-0.41	-0.54	-0.57
Cl	0.98	0.97	1.00	0.99	0.92	0.99	0.97	0.95	0.95	0.78	0.59	0.56	0.28	0.22	0.16	0.01	-0.01	-0.37	-0.54	-0.51
Na	0.97	0.97	0.99	1.00	0.94	0.92	0.98	0.91	0.95	0.81	0.56	0.60	0.23	0.30	0.26	0.01	0.00	-0.37	-0.56	-0.50
Li	0.96	0.95	0.92	0.94	1.00	0.87	0.88	0.82	0.79	0.94	0.72	0.44	0.30	0.53	0.45	0.13	0.29	-0.28	-0.59	-0.49
EC*	0.94	0.91	0.98	0.92	0.87	1.00	0.87	0.92	0.88	0.69	0.71	0.34	0.48	0.02	0.07	0.21	0.09	-0.08	-0.61	-0.69
Br	0.93	0.94	0.97	0.98	0.88	0.87	1.00	0.89	0.96	0.74	0.41	0.72	0.09	0.30	0.29	-0.05	-0.15	-0.43	-0.54	-0.43
Ca	0.97	0.89	0.95	0.91	0.82	0.92	0.89	1.00	0.88	0.65	0.68	0.37	0.47	-0.07	-0.11	0.04	0.03	-0.23	-0.50	-0.51
Sr	0.87	0.88	0.95	0.95	0.79	0.88	0.96	0.88	1.00	0.62	0.36	0.67	0.13	0.08	0.11	-0.10	-0.25	-0.38	-0.52	-0.46
NH_4	0.83	0.83	0.78	0.81	0.94	0.69	0.74	0.65	0.62	1.00	0.65	0.36	0.21	0.77	0.61	0.08	0.41	-0.32	-0.52	-0.41
K	0.71	0.65	0.59	0.56	0.72	0.71	0.41	0.68	0.36	0.65	1.00	-0.23	0.77	0.10	0.01	0.40	0.64	0.15	-0.42	0.71
Mg	0.48	0.53	0.56	0.60	0.44	0.34	0.72	0.37	0.67	0.36	-0.23	1.00	-0.33	0.45	0.37	-0.20	-0.59	-0.58	-0.37	0.03
T*	0.38	0.30	0.28	0.23	0.30	0.48	0.09	0.47	0.13	0.21	0.77	-0.33	1.00	-0.30	-0.39	0.62	0.33	0.42	-0.28	-0.78
$\frac{^{87}Sr}{^{86}Sr}$	0.32	0.36	0.22	0.30	0.53	0.02	0.30	-0.07	0.08	0.77	0.10	0.45	-0.30	1.00	0.86	0.33	0.20	-0.43	-0.36	-0.18
HCO_3	0.23	0.26	0.16	0.26	0.45	0.07	0.29	-0.11	0.11	0.61	0.01	0.37	-0.39	0.86	1.00	0.03	0.27	-0.19	-0.31	0.02
pH*	0.13	0.08	0.01	0.01	0.13	0.21	-0.05	0.04	-0.10	0.08	0.40	-0.20	0.62	0.33	0.03	1.00	0.32	0.57	-0.44	-0.73
$\delta^{13}C$	0.12	0.08	-0.01	0.00	0.29	0.09	-0.15	0.03	-0.25	0.41	0.64	-0.59	0.33	0.20	0.27	0.32	1.00	0.33	-0.04	-0.38
F	-0.32	-0.41	-0.37	-0.37	-0.28	-0.08	-0.43	-0.23	-0.38	-0.32	0.15	-0.58	0.42	-0.43	-0.19	0.57	0.33	1.00	-0.13	-0.14
SO_4	-0.58	-0.54	-0.54	-0.56	-0.59	-0.61	-0.54	-0.50	-0.52	-0.52	-0.42	-0.32	-0.28	-0.36	-0.31	-0.44	-0.04	-0.13	1.00	0.52
$\frac{^3He}{^4He}$	-0.58	-0.57	-0.51	-0.50	-0.49	-0.69	-0.43	-0.51	-0.46	-0.41	-0.71	0.03	-0.78	-0.18	0.02	-0.73	-0.38	-0.14	0.52	1.00

Figure 2: Correlation matrix among the major solute concentrations and physical–chemical properties. Color coding represents the correlation coefficient: large positive values (0.5) are red, intermediate values (0.5 to -0.5) are white, and small values (-0.5) are blue. EC*, electrical conductivity; T*, temperature; pH*, hydrogen ion exponent.

Figure 3 presents the $\delta^{18}O$-δD diagram, in which the data form a linear trend toward a magmatic water composition distinct from the meteoric water line or seawater. Figure 4 presents the selected major element and δD diagrams, again showing linear trends of least squares regression with high coefficients of determination ($R^2 = 0.84$ to 0.99). In Figure 3 and in the bottom two subpanels of Figure 4, each data point is labeled with the (abbreviated) sample number as in Tables 1 and 2, indicating a consistent sample array from AW-2, which is closer to or almost on the meteoric water line, to AW-11, which is closer to magmatic water. It should be noted that the same symbols are used in Figures 2 and 3; light blue diamonds represent the samples from Arima, and dark blue diamonds represent the samples from Takarazuka.

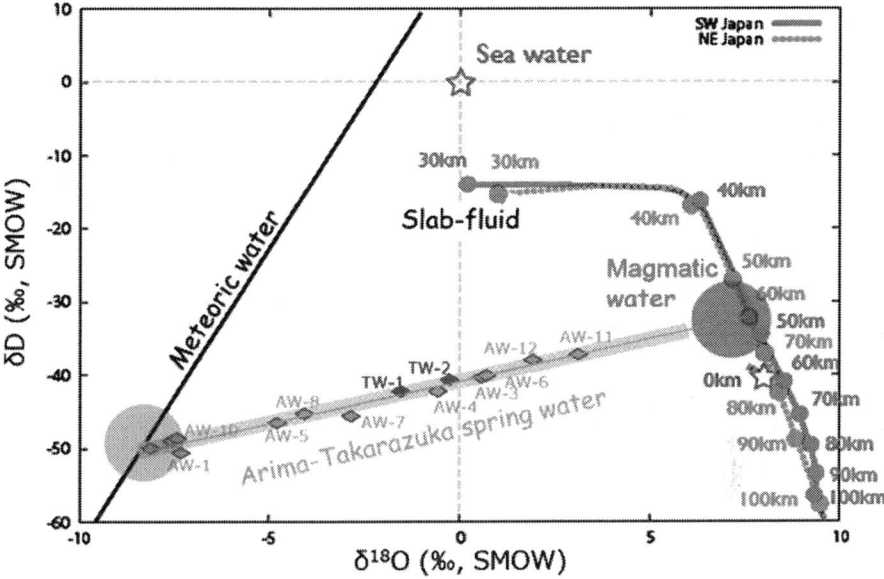

Figure 3: δ18O versus δD diagram of the Arima hot spring waters and the predicted δ18O-δD evolution lines of slab-derived fluids.Red and green lines represent the evolution lines as a function of depth for the Philippine Sea slab (beneath southwestern Japan) and the Pacific slab (beneath northeastern Japan), respectively. The solid circles along the evolution lines indicate depth with intervals of 10 km. Red open star represents δ^{18}O and δD of seawater, which is defined as 0‰ standard mean ocean water (SMOW). Blue open star at δ^{18}O and δD ~ (+8,-40) represents the initial values of altered oceanic crust (AOC) just before subduction (Muehlenbachs 1986; Kawahata et al. 1987). Solid black line in the negative δ^{18}O field represents the average meteoric water line in Japan (Masuda et al.1985). Large purple circle labeled 'Magmatic water' represents the reported range of the worldwide δ^{18}O and δD of magmatic fluids, fumarolic gases, and andesitic waters after White (1957), Matsuo et al. (1974), Sakai and Matsubaya (1977), Mizutani (1978), and Giggenbach (1992).

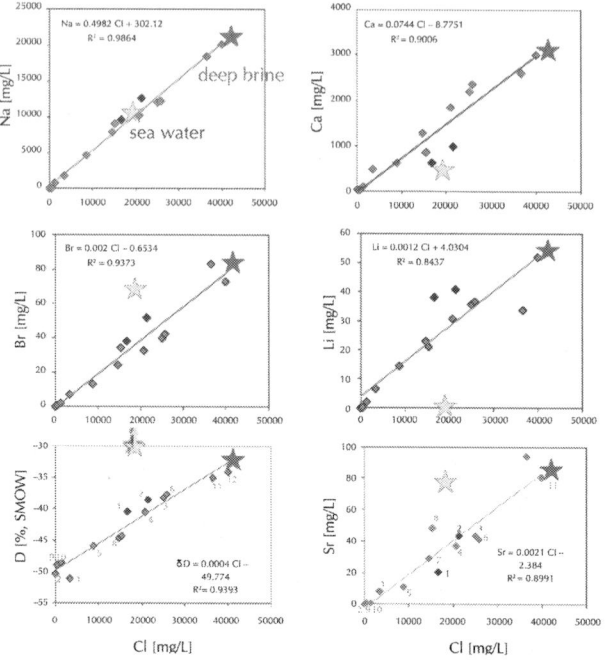

Figure 4: Variation diagrams for chlorine versus major elements and δD for Arima hot spring waters. The least squares fitting line expressed by the equation in each subpanel is shown with the coefficient of determination (R^2).

The new dataset and the linear trends confirm the previous results, which suggest mixing of local meteoric water in the Arima area (δD approximately -50‰ with nearly zero for all the solute concentrations, Masuda et al. 1985) and a deep-seated brine (e.g., Matsubaya et al. 1973; Tanaka et al. 1984; Masuda et al. 1985). Based on the new data, we quantitatively estimated the deep brine composition. The previously observed relationship between ^3H and δD of the Arima hot spring waters, δD of the tritium-free water, i.e., a pure deep brine, is inferred to be -30‰ to -35‰ (Tanaka et al. 1984). Then, the least squares regression lines at δD = -30‰ to -35‰ provide the estimate on the brine composition: $Cl^- = 42,000$ mg/L, $Na^+ = 21,000$ mg/L, $Li^+ = 55$ mg/L, $Br = 84$ mg/L, $Ca^{2+} = 3,100$ mg/L, $Sr^{2+} = 86$ mg/L, $NH_4^+ = 8.8$ mg/L, $K^+ = 3,700$ mg/L, and $δ^{18}O = +6‰$ at δD = -33‰, which are similar to but slightly more diluted than the previous estimates by Masuda et al. (1985).

Oxygen and Hydrogen Isotopes of Slab-Derived Fluid

In order to test the origin of the deep brine, a numerical model was developed to estimate the oxygen and hydrogen isotope compositions of the slab-derived fluid. The model considers the thermal structure along the subducting slab in a 2D across-arc section, stability and dehydration reaction of hydrous mineral phases, and the resultant partitioning of oxygen and hydrogen isotopes between the fluid and the residual mineral phases.

First, the thermal structure of the subduction zone was calculated because dehydration from subducting slabs is controlled thermodynamically and hence chiefly by temperature and pressure. Since dehydration mainly occurs within hydrated oceanic crusts consisting of both sediment and AOC rather than the underlying oceanic mantle (e.g., Iwamori 2007), the temperature distribution along the surface of a slab is calculated as a function of subduction velocity (V), slab age (a), and subduction angle (θ), using an analytical expression based on that reported by England and Wilkins (2004). As a result, we estimated the thermal profile along the slab surface for the subducting Philippine Sea slab ($V=4$ cm/year, $a=15$ Ma, $\theta=20°$) beneath the southwestern Japan arc, as well as that for the Pacific slab ($V=10$ cm/year, $a=120$ Ma, $\theta=30°$) beneath the northeastern Japan arc. Next, slab dehydration and the isotopic fractionation of oxygen and hydrogen were calculated, based on the phase relation of the subducted materials. Although oceanic sediments initially contain 10 to 30 wt% H_2O (Plank and Langmuir 1998), most of water is liberated, decreasing to 2 to 4 wt% by extensive compaction and dehydration of minerals at depths shallower than 30 km (You et al. 1996; Hyndman and Peacock 2003; Kasahara 2003). The sedimentary column currently delivered to deep sea trenches is typically 50 to 500-m thick, and sediment 350-m-thick subducts from the Nankai Trough (Plank and Langmuir 1998). These values are 1/100 to 1/10 thinner than the AOC thickness, limiting its impact on the water budget in subduction zones. Considering these factors, although oceanic sediments still contain 1 to 2 wt% of H_2O at depths greater than 100 km, the AOC-derived fluids are thought to be a major source that supplies deep-seated fluids to the overlying mantle wedge and the arc crust over the slab depth ranging from 30 to 100 km (Iwamori2007), including the Arima area (Figure 1). Therefore, to simplify the model, we considered only dehydration of AOC and its phase relations for dehydration reaction (Schmidt and Poli 1998). It should be noted, however, that several key elements/ isotopes such as carbon and lead, which are particularly concentrated in sediments, may have significant impacts on the fluid compositions (e.g., Sano and Marty 1995; Aizawa et al. 1999; Nakamura et al. 2008), as will be subsequently discussed.

A model was then designed to estimate the isotopic characteristics of slab-derived fluids for the hydrogen and oxygen stable isotopic ratios that constitute the characteristic features of Arima-type brine. The major constituent minerals of AOC at the pressure-temperature condition beneath the Arima area (i.e., 1.5 to 2.4 GPa and 400°C to 500°C, including uncertainty in the estimates), where major dehydration of AOC occurs, are amphibole (22% to 48% in modal composition), lawsonite (22% to 36%), chlorite (0% to 13%), and epidote (9% to 22%; Schmidt and Poli 1998). Accordingly, the following water/mineral isotopic fractionation factors (α) for the major constituent minerals of AOC are used.

Amphibole: The hydrogen isotopic fractionation factor (Graham et al. 1984) is expressed with

$$10^3 \ln\alpha = +23 \ (T < 750°C), \text{ or}$$
$$= 26\left(10^6 T^{-2}\right) - 3 \ (T > 750°C) \tag{1}$$

and the oxygen isotopic fractionation factor (Zheng et al. 1994) follows

$$10^3 \ln\alpha = -0.42\left(10^6 T^{-2}\right) + 2.4 \tag{2}$$

Lawsonite: The hydrogen isotopic fractionation factor (Suzuoki and Epstein 1976) is expressed with

$$10^3 \ln\alpha = +26.4 \ (T < 350°C), \text{ or}$$
$$= 22.4\left(10^6 T^{-2}\right) - 28.2 - 2\left(10^3 T^{-1}\right) \ (T > 350°C) \tag{3}$$

and the oxygen isotopic fractionation factor (Wenner and Taylor 1971) follows

$$10^3 \ln\alpha = -2.3\left(10^6 T^{-2}\right) + 3.8$$

(4)

Chlorite: The hydrogen isotopic fractionation factor (O'Neil 1986) is expressed with

$$10^3 \ln\alpha = +32.7 \ (T < 350°C), \ \text{or}$$
$$= 2.67\left(10^6 T^{-2}\right) + 25.3 \ (T > 350°C)$$

(5)

and the oxygen isotopic fractionation factor (Wenner and Taylor 1971) follows

$$10^3 \ln\alpha = -1.56\left(10^6 T^{-2}\right) + 4.70$$

(6)

where T is the absolute temperature. Epidote is one of the major constituent minerals in AOC at up to 22 wt% (Schmidt and Poli 1998) for the P-T range of interest up to 3 GPa and 600°C in this study, respectively. Matthews et al. (1983) reported that below 600°C, the oxygen exchange reaction is very slow between water and zoisite, the Fe-free end-member of epidote group minerals; therefore, no isotopic fractionation occurs. Instead, dissolution-precipitation appears to occur. In addition, the zoisite-water fractionation factor constrained by mineral-pair equations is around unity (Matthews et al. 1983), causing no $^{18}O/^{16}O$ fractionation. For these reasons, we ignored epidote in our estimates.

The Rayleigh fractionation of oxygen and hydrogen isotopes was calculated by numerically integrating the following equation along the calculated P-T path for the Philippine Sea slab to obtain the compositional path in Figure 3: $\Delta\delta_f/1,000 = -(1-F)/F \ (\alpha_{f/r}-1) \times \ln(1-F)$ (Equation 7), where $\Delta\delta_f$ is the change in isotopic ratio of the fluid corresponding to the infinitesimal change in P-T (‰), F is the extent of dehydration reaction constrained from the phase diagram as a function of P-T (weight fraction), and $\alpha_{f/r}$ is the isotopic fractionation coefficient between the fluid and the bulk rock calculated as $\Sigma X_i \ \alpha_i$, where X_i represents the weight fraction of mineral i, and α_i is the fractionation factor derived from Equation 1 to 6.

The calculation starts with an average composition of AOC ($\delta D = -40$‰, $\delta^{18}O = +8$‰) obtained from a DSDP Hole 504B drill core >1,000-m long that includes fine-grained basalt to dolerite

(Muehlenbachs 1986; Kawahata et al. 1987), as shown by the open star labeled '0 km' in Figure 3. The δD value used (i.e., -40‰) is within the range reported from ODP/IODP Hole 1256D (-64‰ to -25‰, Shilobreeva et al. 2011). A shallow dehydration of this AOC produces relatively light oxygen isotopes ($\delta^{18}O$ of approximately 0‰) and heavy hydrogen isotopes (δD of approximately -15‰) compared with those estimated in the Arima deep brine, which evolves toward the bottom right of the diagram. The results, showing larger $\delta^{18}O$ and smaller δD values as the subduction progresses, reflect continuous dehydration and Rayleigh isotopic fractionation. The subducted young Shikoku Basin and its extinct spreading center (approximately 15 Ma; Okino et al. 1994; Sano et al. 2009) beneath the Arima area may share thermal and hydration conditions with a young oceanic crust at the 504B site off the Costa Rica Rift (approximately 6 Ma), partly justifying the assumed isotopic composition in Figure 3. Nevertheless, subducting slabs exhibit significant compositional variability, e.g., $\delta D = -30$‰ to -60‰, including sediment, AOC, and serpentinite (Kawahata et al. 1987; Giggenbach 1992; Matsuhisa 1992; Shaw et al. 2008), and causes significant uncertainty in the quantitative results of our model. As a consequence, the slab dehydration depth estimated for the Arima deep brine ranges from 45 to 70 km, as shown by the pink circle in Figure 3).

Beneath the Arima area, the depth of the Philippine Sea slab surface is estimated to be 50 to 70 km with relatively large uncertainty (Figure 1; Nakajima and Hasegawa 2007; Hirose et al.2008). Our numerical model suggests that the slab-derived fluid at such depth has an isotopic composition similar to that of geochemically estimated deep brine within the uncertainty, as shown in Figure 3.

The estimated brine composition of δD at approximately -30‰ is similar to that estimated from analysis on back-arc basin basalts or melt inclusions from the Mariana subduction zone such as -25‰ (Poreda 1985) and -32‰ (Shaw et al. 2008), respectively, and the mass balance of subduction zone-scale material cycling at $\delta D = -27$‰ (Kazahaya 1997). The compositional range resembles magmatic fluids, although Arima is far from the volcanoes and is located in the forearc region in terms of northwestward subduction of the Philippine Sea Plate from the Nankai Trough. Westward subduction of the Pacific Plate causes magmatism in central Japan (e.g., Nakamura et al. 2008); Arima is located approximately 300 km west in the backarc region of the magmatism (Figure 1). The current understanding of the magma genesis, including the regional understanding along the entire Japan arcs (Iwamori 2007), indicates no magmatic production or supply beneath the Arima area.

DISCUSSION AND CONCLUSIONS

The ^{3}He/^{4}He ratio may provide crucial information of whether the observed geochemical features reflect the slab-derived fluid directly or if they are affected significantly by arc crust during ascent because the incorporation of crustal components significantly reduces the ratio. Figure 5a shows the ^{3}He/^{4}He ratios for the brine samples as well as the gas samples from the three gas-rich brines. Almost all of the sample plots within the two mixing curves including one between air-saturated water (ASW) and the upper mantle component (8 Ra) and the other between ASW and a component having ^{3}He/^{4}He ratios of 8×10^{-6} (5.7 Ra), indicating high contribution of the mantle component with minor crustal contamination (Figure 5a). Masuda et al. (1985) argued that the high ^{3}He/^{4}He ratios of the Arima hot spring brines are attributed to deep-lying magma and its interaction with the crustal rocks, together with diluted carbonated waters from the basement Paleozoic sedimentary rocks and the Cretaceous to Paleogene acidic igneous rocks that account for the overall $\delta^{18}O$-δD and solute variations of the Arima hot spring waters. Magmatic activity is certainly important for the origin of deep brines in some geothermal fields such as Salton Sea and Cerro Prieto with high Cl content and high ^{3}He/^{4}He ratios, where heat and material input (particularly gas in this case) from subaerial but mid-ocean ridge basalt (MORB)-type magmatism at depth interacts with the overlying thick sedimentary basin (Mazzini et al. 2011; Schmitt et al.2013).

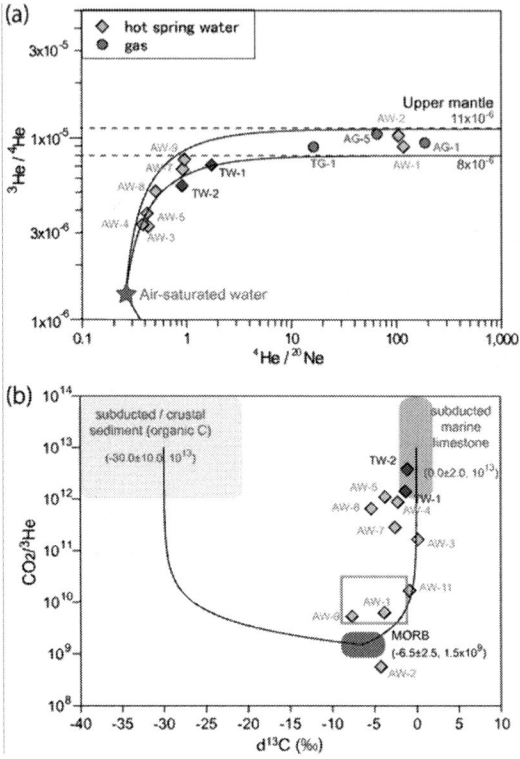

Figure 5: Helium and carbon systematics of the sampled waters and gases. (a) $^{4}He/^{20}He$ versus $^{3}He/^{4}Ne$ diagram of the sampled waters and gases. Light blue diamonds represent samples from Arima; dark blue diamonds (TW-1 and TW-2) represent samples from Takarazuka. Red circles represent gas samples. Two mixing curves are drawn, one between air-saturated water (ASW) and the upper mantle component and the other between ASW and a component having $^{3}He/^{4}He$ ratio of 8×10^{-6}, considering possible but slight crustal contamination. (b) $\delta^{13}C$ versus $CO_2/^{3}He$ diagram of the sampled waters. Yellow, pink, and purple boxes represent the three end-members of sediment-, limestone-, and mid-ocean ridge basalt (MORB)-derived carbon, respectively (Sano and Marty 1995). Solid curves represent mixing between MORB and limestone and MORB and sediment. Open gray box represents the compositional range for arc volcanic gases (Sano and Marty 1995).

However, as discussed in the previous section, the Arima hot spring is different in that no magmatic activity is expected beneath the area and no thick sedimentary basin exists (Kozuki1962). Since the absolute concentrations of ^{3}He in AOC, oceanic sediment, and ASW are four to six orders of magnitude lower than those in MORB (Zelenski et al. 2012), the high ^{3}He/^{4}He ratios found in the Arima hot springs are attributed to mantle helium incorporated in the fluid upon its ascent in the mantle. Arima is located in an area with hot springs of relatively high ^{3}He/^{4}He ratios (greater than 2.5 Ra) known as the Kinki Spot (Sano and Wakita 1985; Figure 1). The Philippine Sea slab subducted beneath the Kinki Spot is young and hot because it is associated with the former spreading ridge (Sano et al. 2009), which may have enhanced the degassing of the mantle.

The ^{3}He/^{4}He ratios are along the mixing line between the air and the mantle component, although the proportions of mixing differ from those identified from the solute concentrations (Figure 4). It is argued that shallow reservoirs exist beneath Arima (e.g., Masuda et al. 1985), where the meteoric water saturated with air mixes with both the deep brine and the exsolved gas and may cause variable mixing of the solutes and the gases. The broad negative correlation between the major solute elements and ^{3}He/^{4}He (Figure 2; Figure 5a) suggests that the spring waters diluted significantly by meteoric water in the shallow reservoir (e.g., AW-1, AW-2, and AW-9, Figures 2and 3) have been fluxed by high ^{3}He/^{4}He gases (e.g., AG-1 and AG-5, Table 2 and Figure 5a) possibly derived from the deep brine (e.g., AW-3, AW-4, and AW-5) that ascends to the surface without being re-gassed.

Carbon dioxide is one of the dominant volatiles in magmatic fluids and gases in subduction zones, which is also one of the distinguishing characteristics of Arima-type brine, and could have originated from slab fluids (Jarrard 2003; Fischer 2008). The plausible sources of carbon are (1) organic carbon in sediments, which are mainly in the continental crust although subducted oceanic sediments are reported to have similar values, (2) carbonate in subducted MORB (AOC), and (3) inorganic carbonate in subducted marine limestones (Nissenbaum et al. 1972; Marty and Jambon1987; Sano and Marty 1995). In order to identify the origin of carbon, the correlation between δ^{13}C and CO_2/^{3}He ratios is useful; in fact, hot spring gases in subduction zones can be explained by three-component mixing, as shown in Figure 5b (Sano and Marty 1995). Within this context, the observed carbon isotopic compositions of the Arima hot spring waters also suggest possible involvement of deep-seated slab materials such as subducted

MORB and limestone rather than organic carbon from a shallow sedimentary reservoir (Figure 5b). The poor correlation between the major solute, mostly derived from subducted MORB, and the carbon isotopic ratios, mostly buffered by subducted limestone, may reflect decoupled processes of different lithologies in a subducted slab such as dehydration versus decarbonation.

The $CO_2/^3He$ ratios vary significantly over four orders of magnitude with broadly constant $\delta^{13}C$ (Figure 5b). Low $CO_2/^3He$ samples (AW-1, AW-2, and AW-9; Figure 5b) exhibited relatively high $^3He/^4He$ (Figure 5a) and low solute concentrations (Figure 4), when compared with the high $CO_2/^3He$ samples (AW-3, AW-4, and AW-5) with low $^3He/^4He$ and high solute concentrations. Sano and Marty (1995) argued that the $CO_2/^3He$ ratio of bubbling gas is lower than that of volcanic hydrothermal water which exsolves the gas because CO_2 is much more soluble to water than He. They suggested that significant $CO_2/^3He$ fractionation occurs during the ascent of deep hydrothermal fluids associated with decompressional degassing. Such a fractionation mechanism may consistently explain the lower $CO_2/^3He$ samples (AW-1, AW-2, AW-9) that have been fluxed by the gas with low $CO_2/^3He$ and high $^3He/^4He$ ratios in a meteoric water-dominated shallow reservoir, in addition to the high $CO_2/^3He$ samples (AW-3, AW-4, AW-5) that contain more of the deep brine component from which the gas has been exsolved.

It has been argued that a high concentration of Cl^- indicates an extensive water-rock interaction either at depth of a high temperature or with the granitic basement rocks at a relatively shallow depth (Edmunds et al. 1985). In the latter case, Cl^- is derived from granitic rocks by acid hydrolysis of biotite and plagioclase. Since hydrolysis of biotite could generate Cl^-, Li^+, K^+, and other species in the groundwater, Cl^- and Li^+ can be considered as conservative products of biotite alteration. On the contrary, the acid hydrolysis of plagioclase may contribute to the principal sources of Na^+ and Ca^{2+} in the groundwater without supplying Cl^- or Li^+. Molar Cl^-/Li^+ and Na^+/Ca^{2+} ratios of the sampled waters, however, reach up to about 150 and 12, respectively, which are significantly larger than the stoichiometry deduced from the above reactions. Although the possibility of such an excess of Na^+ and Cl^- owing to decreases in Ca^{2+}, K^+, and Mg^{2+} through the formation of clay minerals in the aquifer cannot be entirely ruled out, it appears difficult to attain the high salinity (approximately 42,000 ppm Cl) solely by chemical interactions between the basement granites and water (Edmunds et al. 1985).

The concentrations and isotopic ratios of Sr support binary mixing between diluted near-surface water (approximately represented by sample AW-2) and high-Sr deep brine (represented by AW-12; Figure 6). The estimated $^{87}Sr/^{86}Sr$ ratio of the deep brine is close to the estimated ratio of the Philippine Sea slab-derived fluid at 0.71087 and is distinct from the Pacific slab-derived fluid at 0.70538, possibly because the former has more terrigenous components subducted from the Nankai Trough (Plank and Langmuir 1998; Nakamura et al. 2008). The brine composition is also close to the lowermost range of the basement rocks including Rokko granite and the Arima group (approximately 0.707 to 0.710; Terakado and Nohda 1993). The interactions between the solutions and the basement rocks may have, to some extent, perturbed the $^{87}Sr/^{86}Sr$ ratios of the brine, which could have caused the poor correlation between the major solute concentrations (including Sr) and $^{87}Sr/^{86}Sr$. However, the arguments above in terms of helium, carbon, and major solute stoichiometry suggest that the one end-member of the trend is likely to be deep brine originating from the subducted Philippine Sea slab, rather than the host country rocks.

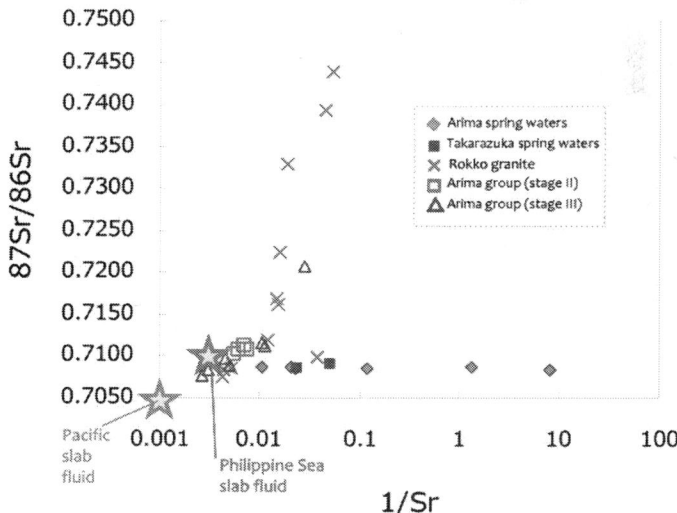

Figure 6: Analyzed Sr concentration expressed by 1/Sr (ppm-1) versus 87Sr/86Sr of the Arima-type brine. Sr concentration expressed by 1/Sr

(ppm^{-1}) versus $^{87}Sr/^{86}Sr$ of the Arima-type brine analyzed in this study, together with the data of the Rokko granitoids and the Arima group consisting mainly of rhyolitic welded tuffs (Terakado and Nohda 1993). Large red and green stars represent slab fluid compositions derived from the subducted Philippine Sea Plate and Pacific Plate, respectively (Nakamura and Iwamori 2009). The analytical errors of $^{87}Sr/^{86}Sr$ ratio in this study are approximately 3×10^{-6}.

Numerical forward modeling in this study suggests that slab fluids with $\delta^{18}O$-δD values similar to the Arima-type brine are produced along the subducted Pacific slab beneath the northeastern Japan arc (Figure 3). However, the produced fluid reacts with the mantle wedge just above the slab to form serpentinite due to the cold geotherm along the slab (Iwamori 2007). Therefore, the fluid is absorbed by the mantle and subducted to the deeper part along the slab, which may explain the absence of the Arima-type brine, to our knowledge, in the non-volcanic forearc region of northeastern Japan arc. On the contrary, in a warmer environment beneath the southwestern Japan arc, such an 'absorption' effect by serpentinite is greatly reduced (Iwamori 2007), and the slab fluid may escape to the surface without significant reaction with the country rocks, possibly guided by large faults at crustal levels.

It is also noted that the amount of fluid generated from the slab is maximum at depths of 30 to 80 km and is greatly reduced deeper than 100 km in most subduction zones (Iwamori 1998). Instead, beneath the northeastern Japan arc, the subducted serpentinite along the slab undergoes dehydration and supplies fluids upward beneath the arc volcanic zones, causing melting in a high-temperature region of the central part of the mantle wedge (Iwamori 1998). Then, hydrous magma is fed to a shallow chamber, where degassing may create volcanic waters and gases. Therefore, $\delta^{18}O$-δD compositions of the volcanic thermal waters, widely known as 'andesitic water' having $\delta^{18}O = +5‰$ to $+10‰$ and $\delta D = -10‰$ to $-30‰$ (Taran et al. 1989; Giggenbach 1992), integrate and reflect these processes of slab dehydration, sepentinite formation-dehydration, reaction with the mantle to cause melting, and degassing, as discussed by Matsuhisa (1992). In this context, andesitic water and the Arima-type brine as suggested in this study are believed to reflect different origins, although their $\delta^{18}O$-δD ranges overlap.

The estimated solute concentrations in the Arima deep brine (e.g., 42,000 ppm Cl; 3,700 ppm K; 84 ppm Br; $Br/Cl \sim 2.0 \times 10^{-3}$) are consistent with the recent knowledge on slab-derived fluids, e.g., 12,000 to 240,000 ppm Cl, 600 to 3,700 ppm for K, $Br/Cl \sim 1.0$ to 3.7×10^{-3} for fluids derived from subducted serpentinite (Kendrick et al. 2011), and 5.1 wt% NaCl (approximately 30,000 ppm Cl) for fluid inclusions in the mantle xenoliths (Kawamoto et al. 2013). On the contrary, the Philippine Sea slab-derived fluid compositions estimated from volcanic rocks in central Japan (Nakamura and Iwamori 2013) have higher concentrations for K and Sr (64,000 and 268 ppm, respectively), possibly reflecting different dehydration conditions of higher pressure and temperature beneath the volcanic region compared with those in the forearc region in southwestern Japan (Nakamura et al. 2014). It is also noted that during deep processes of dehydration and melting in subduction zones, Cl and F are reported to behave very differently (Straub and Layne 2003), which may explain the poor correlation between the major solute including Cl and F (Figure 2).

In any case, the composition of the Arima brine appears to retain the deep fluid signatures derived from multiple sources such as the subducted Philippine Sea slab and the overlying mantle, including (i) AOC for most water and hence hydrogen and oxygen isotopes in addition to associated elements such as Cl, Na, Li, Br, Ca, and Sr (Figure 2), (ii) subducted sediment, particularly limestone, for carbon, and (iii) mantle wedge for a high $^3He/^4He$ component. These signatures indicate that deep-seated fluid may reach the surface without significant interaction during its ascent, such as through channel flow from depth, as also suggested from the seismic velocity variations (Iwamori and Nakakuki 2013) and trace element transport in subduction zones (Ikemoto and Iwamori 2014). However, only a small amount of contamination by the basement granitic rocks with very high $^{87}Sr/^{86}Sr$ may have affected the Sr isotopic ratio of the fluid, and some crustal helium with low $^3He/^4He$ could have affected the degassed water samples such as AW-3, AW-4, and AW-5 with relatively low He content (Table 2). If such a fracture system is a dominant mode of fluid transport, the Arima-type brine may put direct constraints on both the compositions and processes of slab fluid at depth.

AUTHORS' CONTRIBUTIONS

CK, HI, HN and KK designed the study, CK and HI sampled the waters and gases, CK and NM carried out the elemental and isotopic analyses, CK and HI constructed and performed numerical modeling, all authors discussed the data to make interpretation, wrote the text, prepared the figures and tables, and read and approved the final manuscript.

ACKNOWLEDGEMENTS

The authors would like to thank HA Takahashi, M Takahashi, M Tanimizu, and T Ishikawa for their help in elemental and isotopic analyses, M Totani for his kind arrangement and assistance in sampling hot spring waters in the Arima area, and the two anonymous referees for constructive review that greatly improved the manuscript.

REFERENCES

1. Aizawa Y, Tatsumi Y, Yamada H (1999) Element transport by dehydration of subducted sediments: implication for arc and ocean island magmatism. Island Arc 8:38-46

2. Arai T (2007) Geochemical characteristics of whole rock and minerals from the Late Cretaceous granitic rocks of the Rokko Mountains. J Mineral Petrol Sci 102:12-23

3. Birck JL (1986) Precision K–Rb–Sr isotopic analysis: application to Rb–Sr chronology. Chem Geol 56:73-83

4. Brenan JM, Shaw HF, Ryerson FJ, Phinney DL (1995) Mineral-aqueous fluid partitioning of trace elements at 900_C and 2.0 GPa: constraints on the trace element chemistry of mantle and deep crustal fluids. Geochim Cosmochim Acta 59:3331-3350

5. Edmunds WM, Kay RLF, McCartney RA (1985) Origin of saline groundwaters in the Carnmenellis granite (Cornwall, England): natural processes and reaction during hot dry rock reservoir circulation. Chem Geol 49:287-301

6. England P, Wilkins C (2004) A simple analytical approximation to the temperature structure in subduction zones. Geophys J Int 159:1138-1154

7. Fischer TP (2008) Fluxes of volatiles (H_2O, CO_2, N_2, Cl, F) from arc volcanoes. Geochem J 42:21-38

8. Giggenbach WF (1992) Isotopic shifts in waters from geothermal and volcanic systems along convergent plate boundaries and their origin. Earth Planet Sci Lett 113:495-510

9. Graham CM, Harmon RS, Sheppard SMF (1984) Experimental hydrogen isotope studies—hydrogen isotope exchange between amphibole and water. Am Miner 69:128-138

10. Hasegawa A, Nakajima J, Kita S, Tsuji Y, Nii K, Okada T, Matsuzawa T, Zhao D (2008) Transportation of H_2O in the NE Japan subduction zone as inferred from seismic observations. J Geophys Res 117:59-75 (in Japanese with English abstract)

11. Hirose F, Nakajima JI, Hasegawa A (2008) Three-dimensional seismic velocity structure and configuration of the Philippine Sea slab in southwestern Japan estimated by double-difference tomography. J Geophys Res 113: B09315, doi: 10.1029/2007JB005274

12. Hyndman RD, Peacock SM (2003) Serpentinization of the forearc mantle. Earth Planet Sci Lett 212:417-432

13. Ikemoto A, Iwamori H (2014) Numerical modeling of trace element transportation in subduction zones: implications for geofluid processes. Earth Planets Space 66:26 doi: 10.1186/1880-5981-66-26

14. Ionov DA (2010) Petrology of mantle wedge lithosphere: new data on supra-subduction zone peridotite xenoliths from the Andesitic Avacha Volcano, Kamchatka. J Petrol 51:327-361

15. Ishikawa T, Nakamura E (1992) Origin of the slab component in arc lavas from across-arc variation of B and Pb isotopes. Nature 370:205-208

16. Ishimaru S, Arai S (2008) Arsenide in a metasomatized peridotite xenolith as a constraint on arsenic behavior in the mantle wedge. Am Mineral 93:1061-1065

17. Iwamori H (1998) Transportation of H_2O and melting in subduction zones. Earth Planet Sci Lett 160:65-80

18. Iwamori H (2007) Transportation of H_2O beneath the Japan arcs and its implications for global water circulation. Chem Geol 239:182-198

19.   Iwamori H, Albarède F (2008) Decoupled isotopic record of ridge and subduction zone processes in oceanic basalts by independent component analysis. Geochem Geophys Geosys 9: doi: 10. 2007GC001753

20.   Iwamori H, Nakakuki T (2013) Fluid processes in subduction zones and water transport to the deep mantle. In: Karato S (ed) Physics and chemistry of the deep earth, Amsterdam: Elsevier. pp 446-468

21.   Jarrard RD (2003) Subduction fluxes of water, carbon dioxide, and potassium. Geochem Geophys Geosys 4:8905 doi: 10.1029/2002GC000392

22.   Kasahara J (2003) The role of water in earthquake generation. In: Kawamura K, Kasahara J (eds) Toriumi M, Tokyo: University of Tokyo Press.

23.   Kawahata H, Kusakabe M, Kikuchi Y (1987) Strontium, oxygen, and hydrogen isotope geochemistry of hydrothermally altered and weathered rocks in DSDP Hole 504B, Costa Rica Rift. Earth Planet Sci Lett 85:343-355

24.   Kawamoto T, Yoshikawa M, Kumagai Y, Mirabueno MHT, Okuno M, Kobayashi T (2013) Mantle wedge infiltrated with saline fluids from dehydration and decarbonation of subducting slab. Proc Natl Acad Sci U S A 110:9663-9668

25.   Kazahaya K (1997) Discharged H_2O from active island arc volcanoes: origin and flux. J Jpn Assoc Hydrol Sci 27:105-116

26.   Kendrick MA, Scambelluri M, Honda M, Phillips D (2011) High abundances of noble gas and chlorine delivered to the mantle by serpentinite subduction. Nature Geo 4: doi: 10.1038/NGEO1270

27.   Keppler H (1996) Constraints from partitioning experiments on the composition of subduction-zone fluids. Nature 380:237-240

28.   Kessel R, Schmidt MW, Ulmer P, Pettke T (2005) Trace element signature of subduction-zone fluids, melts and supercritical liquids at 120–180 km depth. Nature 437:724-727 doi: 10.1038/ nature03971

29.   Kimura JI, Hacker BR, van Keken PE, Kawabata H, Yoshida T, Stern RJ (2009) Arc Basalt Simulator version 2, a simulation for slab dehydration and fluid-fluxed mantle melting for arc basalts: modeling scheme and application. Geochem Geophys Geosys 10: Q09004, doi: 10.1029/2008GC002217

30. Kogiso T, Tatsumi Y, Nakano S (1997) Trace element transport during dehydration processes in the subducted oceanic crust: 1. Experiments and implications for the origin of ocean island basalts. Earth Planet Sci Lett 148:193-205

31. Kozuki J (1962) Studies of Arima Spa. Nippon Shoin, Tokyo. (in Japanese)

32. Marty B, Jambon A (1987) C/^{3}He in volatile fluxes from the solid Earth: implications for carbon geodynamics. Earth Planet Sci Lett 83:16-26

33. Masuda H, Sakai H, Chiba H, Tsurumaki M (1985) Geochemical characteristics of $Na-Ca-Cl-HCO_3$type waters in Arima and its vicinity in the western Kinki district, Japan. Geochem J 19:149-162

34. Matsubaya O (1981) Origin of hot spring waters based on hydrogen and oxygen isotopic ratios. Hot Spring Sci 31:47-56 (in Japanese)

35. Matsubaya O, Sakai H, Kusachi I, Satake H (1973) Hydrogen and oxygen isotopic ratios and major element chemistry of Japanese thermal water systems. Geochem J 7:123-151

36. Matsuhisa Y (1992) Origin of magmatic waters in subduction zones: stable isotopic constraints. Rept Geol Surv Japan 279:104-109

37. Matsumoto T, Kawabata T, Matsuda J, Yamamoto K, Mimura K (2003) ^{3}He/^{4}He ratios in well gases in the Kinki district, SW Japan: surface appearance of slab-derived fluids in a non-volcanic area in Kii Peninsula. Earth Planet Sci Lett 216:211-230

38. Matsuo S, Suzuoki T, Kusakabe M, Wada H, Suzuki M (1974) Isotopic and chemical compositions of volcanic gases from Satsuma-Iwojima, Japan. Geochem J 8:165-173

39. Matthews A, Goldsmith JR, Clayton RN (1983) Oxygen isotope fractionation between zoisite and water. Geochim Cosmochim Acta 47:645-654

40. Mazzini A, Svensen H, Etiope G, Onderdonk N, Banks D (2011) Fluid origin, gas fluxes and plumbing system in the sediment-hosted Salton Sea Geothermal System (California, USA). J Volcanol Geotherm Res 205:67-83

41. Mizutani Y (1978) Isotopic composition of volcanic steam from Showashinzan Volcano, Hokkaido, Japan. Geochem J 12:57-63

42. Morikawa N, Kazahaya K, Masuda H, Ohwada M, Nakama A, Nagao K, Sumino H (2008) Relationship between geological structure and helium isotopes in deep groundwater from the Osaka Basin: application to deep groundwater hydrology. Geochem J 42:61-74

43. Muehlenbachs K (1986) Alternation of the oceanic crust and the ^{18}O history of seawater. Rev Mineral 16:425-444

44. Nagao K, Takaoka N, Matsubayashi O (1981) Rare gas isotopic compositions in natural gases in Japan. Earth Planet Sci Lett 53:175-188

45. Nakajima JI, Hasegawa A (2007) Subduction of the Philippine Sea slab beneath southwestern Japan: slab geometry and its relationship to arc magmatism. J Geophys Res 112: B08306, doi: 10.1029/2006JB004770

46. Nakamura H, Iwamori H (2009) Contribution of slab fluid in arc magmas beneath the Japan arcs. Gondwana Res 16:431-445 doi: 10.1016/j.gr.2009.05.004

47. Nakamura H, Iwamori H (2013) Generation of adakites in a cold subduction zone due to double subducting plates. Contrib Mineral Petrol 165:1107-1134 doi: 10.1007/s00410-013-0850-0

48. Nakamura H, Iwamori H, Kimura JI (2008) Geochemical evidence for enhanced fluid flux due to overlapping subducting plates. Nature Geo 1:380-384

49. Nakamura H, Fujita Y, Nakai S, Yokoyama T, Iwamori H (2014) Rare earth elements and Sr–Nd–Pb isotopic analyses of the Arima hot spring waters, Southwest Japan: implications for origin of the Arima-type brine. J Geol Geosci 3:161 doi: 10.4172/2329-6755.1000161

50. Nishimura S (2000) Forearc volcanism and hot-springs in Kii Peninsula, Southwest Japan. Hot Spring Sci 49:207-216 (in Japanese with English abstract)

51. Nissenbaum A, Presley BJ, Kaplan IR (1972) Early diagenesis in a reducing fjord, Saanich Inlet, British Columbia - I. Chemical and isotopic changes in major components of interstitial water. Geochim Cosmochim Acta 36:1007-1104

52. O'Neil JR (1986) Stable isotopes in high temperature geological processes. Rev Mineral 16:1-40

53. Okada T, Itaya T, Sato M, Nagao K (1994) Noble gas isotopic composition of deep underground water in Osaka plain, central Japan: evidence for mantle He and model for new volcanism. Island Arc 3:221-231

54. Okino K, Shimakawa Y, Nagaoka S (1994) Evolution of the Shikoku Basin. J Geomagn Geoelectr 46:463-479

55. Pearce JA, Stern RJ, Bloomer SH, Fryer P (2005) Geochemical mapping of the Mariana arc-basin system: implications for the nature and distribution of subduction components. Geochem Geophys Geosys 6: Q07006, doi: 10.1029/2004GC000895

56. Plank T, Langmuir CH (1998) The chemical composition of subducting sediment and its consequences for the crust and mantle. Chem Geol 145:325-394

57. Poreda R (1985) Herium-3 and deuterium in back-arc basalts: Lau Basin and the Mariana Trough. Earth Planet Sci Lett 73:244-254

58. Sakai H, Matsubaya O (1977) Stable isotopic studies of Japanese geothermal systems. Geothermics 5:97-124

59. Sakai H, Matsuhisa Y (1996) Stable isotope geochemistry. University of Tokyo Press, Tokyo. 403 pp; (in Japanese)

60. Sano Y, Marty B (1995) Origin of carbon in fumarolic gas from island arcs. Chem Geol 119:265-274

61. Sano Y, Wakita H (1985) Geographical distribution of $^3He/^4He$ ratios in Japan - implications for arc tectonics and incipient magmatism. J Geophys Res 90:8729-8741

62. Sano Y, Kameda A, Takahata N, Yamamoto J, Nakajima J (2009) Tracing extinct spreading center in SW Japan by helium-3 emanation. Chem Geol 266:50-56doi: 10.1016/j.chemgeo.2008.10.020

63. Schiano P, Clocchiatti R, Shimizu N, Maury RC, Jochum KP, Hofmann AW (1995) Hydrous silica-rich melts in the sub-arc mantle and their relationship with erupted arc lavas. Nature 377:595-600

64. Schmidt MW, Poli S (1998) Experimentally based water budgets for dehydrating slabs and consequences for arc magma generation. Earth Planet Sci Lett 163:361-379

65. Schmitt AK, Martín A, Weber B, Stockli DF, Zou H, Shen C-C (2013) Oceanic magmatism in sedimentary basins of the northern Gulf of California rift. Geol Soc Am Bull 125:1833-1850 doi: 10.1130/B30787.1

66. Shaw AM, Hauri EH, Fischer TP, Hilton DR, Kelley KA (2008) Hydrogen isotopes in Mariana arc melt inclusions: implications for subduction dehydration and the deep-Earth water cycle. Earth Planet Sci Lett 275:138-145

67. Shilobreeva S, Martinez I, Busigny V, Agrinier P, Laverne C (2011) Insights into C and H storage in the altered oceanic crust: results from ODP/IODP Hole 1256D. Geochim Cosmochim Acta 75:2237-2255

68. Shinohara H (2013) Volatile flux from subduction zone volcanoes: insights from a detailed evaluation of the fluxes from volcanoes in Japan. J Volcanol Geotherm Res 268:46-63

69. Straub SM, Layne GD (2003) The systematics of chlorine, fluorine, and water in Izu arc front volcanic rocks: implications for volatile recycling in subduction zones. Geochim Cosmochim Acta 67:4179-4203

70. Suzuoki T, Epstein S (1976) Hydrogen isotope fractionation between OH-bearing minerals and water. Geochim Cosmochim Acta 40:1229-1240

71. Takahashi HA, Nakamura T, Tsukamoto H, Kazahaya K, Handa H, Hirota A (2013) Radio carbon dating of groundwater in granite fractures in Abukuma Province, northeast Japan. Radio Carbon 55:894-904

72. Tanaka K, Koizumi M, Seki R, Ikeda N (1984) Geochemical study of Arima hot-spring waters, Hyogo, Japan, by means of tritium and deuterium. Geochem J 18:173-180

73. Taran YA, Pokrovsky BG, Dubik YM (1989) Isotopic composition and the origin of water from andesitic magmas. Trans Acad Sci USSR 304:440-443

74. Taran YA, Varley NR, Inguaggiato S, Cienfuegos E (2010) Geochemistry of H_2- and CH_4-enriched hydrothermal fluids of

Socorro Island, Revillagigedo Archipelago, Mexico. Evidence for serpentinization and abiogenic methane. Geofluids 10:542-555 doi: 10.1111/j.1468-8123.2010.00314.x

75. Terakado Y, Nohda S (1993) Rb-Sr dating of acidic rocks from the middle part of the Inner Zone of southwest Japan: tectonic implications for the migration of the Cretaceous to Paleogene igneous activity. Chem Geol 109:69-87

76. Umeda K, Ogawa Y, Asamori K, Oikawa T (2006) Aqueous fluids derived from a subducting slab: observed high ^{3}He emanation and conductive anomaly in a nonvolcanic region, Kii Peninsula southwest Japan. J Volcanol Geotherm Res 149:47-61

77. Wenner DB, Taylor HP (1971) Temperatures of serpentinization of ultramafic rocks based on O^{18}/O^{16} fractionation between coexisting serpentine and magnetite. Contrib Mineral Petrol 32:165-185

78. White DE (1957) Magmatic, connate, and metamorphic waters. Geol Soc Am Bull 68:1659-1682

79. You CF, Castillo PR, Gieskes JM, Chan LH, Spivack AJ (1996) Trace element behavior in hydrothermal experiments: implications for fluid processes at shallow depths in subduction zones. Earth Planet Sci Lett 140:41-52

80. Zelenski ME, Taran YA, Dubinina EO, Shapar VN, Polyntseva EA (2012) Sources of volatiles for a subduction zone volcano: Mutnovsky Volcano, Kamchatka. Geochem Int 50:502-521

81. Zheng YF, Metz P, Satir M (1994) Oxygen-isotope fractionation between calcite and tremolite—an experimental study. Contrib Mineral Petrol 118:249-255

Bioprospecting the Thermal Waters of the Roman Baths: Isolation of Oleaginous Species and Analysis of the FAME Profile for Biodiesel Production

Holly D Smith-Bädorf[1], Christopher J Chuck[2], Kirsty R Mokebo[3], Heather MacDonald[4], Matthew G Davidson[3], and Rod J Scott[1]

[1]Department of Biology and Biochemistry, University of Bath, Bath, BA2 7AY, UK

[2]Department of Chemical Engineering, University of Bath, Bath, BA2 7AY, UK

[3]Department of Chemistry, University of Bath, Bath, BA2 7AY, UK

[4]Department of Applied Sciences, University of the West of England, Bristol, BS16 1QY, UK

ABSTRACT

The extensive diversity of microalgae provides an opportunity to undertake bioprospecting for species possessing features suited to commercial scale cultivation. The outdoor cultivation of microalgae is subject to extreme temperature fluctuations; temperature tolerant microalgae would help mitigate this problem. The waters of the Roman Baths, which have a temperature range between 39°C and 46°C, were sampled for microalgae. A total of 3 green algae, 1 diatom and 4 cyanobacterial species were successfully isolated into 'unialgal' culture. Four isolates were filamentous, which could prove advantageous for low energy dewatering of cultures using filtration.

Lipid content, profiles and growth rates of the isolates were examined at temperatures of 20, 30, 40°C, with and without nitrogen starvation and compared against the oil producing green algal species, *Chlorella emersonii*. Some isolates synthesized high levels of lipids, however, all were most productive at temperatures lower than those of the Roman Baths. The eukaryotic algae accumulated a range of saturated and polyunsaturated FAMEs and all isolates generally showed higher lipid accumulation under nitrogen deficient conditions (*Klebsormidium* sp. increasing from 1.9% to 16.0% and *Hantzschia* sp. from 31.9 to 40.5%). The cyanobacteria typically accumulated a narrower range of FAMEs that were mostly saturated, but were capable of accumulating a larger quantity of lipid as a proportion of dry weight (*M. laminosus*, 37.8% fully saturated FAMEs). The maximum productivity of all the isolates was not determined in the current work and will require further effort to optimise key variables such as light intensity and media composition.

INTRODUCTION

'The transport fuel sector constitutes a large proportion of global energy demand but renewable alternatives, in comparison to other energy sectors, are largely underdeveloped (Schenk et al.2008). To be competitive renewable liquid fuels must match the performance, energy density and versatility of current fossil fuels (Rupprecht 2009). It is therefore imperative that renewable, energy dense liquid fuels capable of integration into the existing infrastructure are rapidly developed.

One renewable liquid fuel is biodiesel, comprised of the fatty acid methyl esters (FAMEs) produced from the esterification of biologically derived oils. Biodiesel contains chain lengths between C_{14}-C_{24} with varying degrees of unsaturation (Varfolomeev and Wasserman, 2011). The FAME profile is also dependent on the specific producing organism as well as its growing conditions (Saraf and Thomas, 2007). Thus, biodiesel tends to have variable fuel properties that can substantially affect engine performance (Fortman et al. 2008). Biodiesel contains a relatively high oxygen content by weight which results in more complete combustion than mineral diesel resulting in lower CO, particulate matter and hydrocarbon emissions (Song et al., 2008). Currently biodiesel is produced from agricultural crops such as rapeseed, soybean or palm that occupy limited agricultural resources, are produced on a seasonal basis and in most cases generate a low energy return (Rittmann, 2008). They also pose a threat to biodiversity, as marginal land is often ecologically valuable, constituting habitat or acting as a watershed (Gressel, 2008).

In an attempt to replace these first generation feedstocks, there has been a resurgence of interest in developing oleaginous microalgae. Algal biofuels have potentially high energy returns and a small ecological footprint (Groom et al. 2008), though the FAME profile is reportedly more variable than those produced by terrestrial crops (Haik et al. 2011). Microalgae have the potential for all year round harvest (Chisti, 2008), are lipid-rich and offer the prospect of harnessing the residual biomass in conventional biomass technologies (Um and Kim, 2009). Algae culture can be coupled to industrial CO_2 sequestration (Chisti, 2007), bioremediation or scrubbing of waste streams (Pittman et al. 2011). In addition, algae are capable of producing a multitude of other high-value products including biopolymers, vitamins, antioxidants, polysaccharides, proteins, and pharmaceuticals (Raja et al., 2008). Economic analysis demonstrates that co-production of these products is currently essential for commercial viability (Ono and Cuello, 2006).

Algae are responsible for 50% of global CO_2 fixation and have colonised diverse ecological habitats from oceans to hot springs, snowfields and waste streams (Croft et al. 2006). Microalgae are best described as unicellular, microscopic (2-200μm) photosynthetic organisms as they have both prokaryotic (cyanobacteria) and eukaryotic representatives, with over 300,000 species (Pulz and Gross, 2004). The most abundant microalgae in any given environment are generally the diatoms, green algae, cyanobacteria and golden algae (Mutanda et al. 2011).

Microalgae which can accumulate lipids at over 20% of their dry weight are referred to as oleaginous species, most of which belong to the green algae and diatoms (Pulz and Gross, 2004). Cyanobacteria are also capable of accumulating lipid, yet are comparatively understudied with regard to lipid content and profile (Karatay and Dönmez 2011). *Chlorella* spp. is a commonly studied oleaginous genera with a very high CO_2 fixation rate and lipid content compared to other green algae. In the long term, genetic modification might well have the biggest impact on realising fuels from microalgae but algal bioengineering is in its infancy (Chisti and Yan, 2011; Purton and Stevens, 1997). In addition, transgenic cells usually exhibit lower fitness as well as there being restrictions and public concern over their commercial use, therefore, genetic modification must 'compliment and not substitute screening of new species' (Pulz and Gross, 2004).

Since microalgae are a diverse group of organisms, with large variations in growth requirements, growth rate, fatty acid content and the ability to produce other valuable products, the screening and isolation of new algal species is an important goal in developing algal biofuels (Hu et al. 2008). Choice of species is an essential consideration for any applied microalgal biotechnology (Ratha and Prasanna, 2012). The dilute nature of microalgal cultures means extensive dewatering is required prior to biomass processing, which significantly raises energy costs using current methods such as centrifugation, electrophoresis and floatation (Chisti and Yan, 2011). Harvesting costs are therefore an important consideration when screening for new strains, as differences in morphology can have a substantial impact on the cost of production (Mutanda et al. 2011). Whilst low energy methods such as filtration are challenging with unicellular microalgal species these can be very effective for filamentous species (Uduman et al. 2010 and Mohan et al. 2010).

Bioprospecting of diverse aquatic environments offers a way to obtain new oleaginous microalgae with features suited to commercialisation. Hypersaline and thermophillic springs have proved ideal environments for the identification novel microalgae (Mutanda et al. 2011). Extremophilic algae may be beneficial for industrial applications, as the unique environment minimizes contamination risk, particularly in open pond systems, and can provide buffering against fluctuations in temperature (Pulz and Gross, 2004). Significantly, the most productive commercially cultivated species could be considered extremophillic (Xu et al. 2009).

Commonly cultured microalgal species have optimal growth between 16-27°C, with temperatures below 16°C resulting in slow growth, and temperatures above 35°C often being lethal (Barsanti and Gualtieri, 2006). Although thermotolerant microalgal species have been isolated from hot springs (primarily in the USA and Japan), few have been extensively studied to date (Ratha and Prasanna, 2012).

The largest geothermal spring in the UK is situated in the city of Bath at the Roman Baths (ST750647) (Atkinson and Davison, 2002). A series of 3 thermal springs first supply water at a rate 720 l h^{-1} to the Kings Bath (46.5°C), which then drains into the Great Bath 39.0°C; (Andrews et al. 1982). The baths were gradually constructed by Roman settlers between 70-370 AD, but fell into disrepair gradually becoming buried during the 5th century as the Romans withdrew from Britain (pers. comm. Tom Byrne). Excavation of the Baths began in 1878 and have remained open to the elements for the last 130 years (Kellaway, 1991). Since thermotolerance is attractive for biofuel production we set out to bioprospect the hot waters of the Roman Baths for microalgal strains with an ability to produce lipids for biodiesel production.

MATERIALS AND METHODS

Sampling of Bath Water for Algae

Samples of water (1l) were taken from the Great Bath and the Kings Bath and analysed by Severn Trent Services, the pH and temperature were measured on site. Three visible filamentous algae were sampled from scrapings taken from microbial mats submerged beneath the water surface. Approximately 90 l of water from the Great Bath was filtered through a custom-made filter-stack at 3 positions around the Great Bath (the water entry point, exit point and opposite). The custom filter stack comprised of an acrylic tube with a coarse 3 mm mesh metal filter, a sponge filter and a Whatman GF fine filter paper. The GF filter paper layer was used as a sample for further culturing.

Establishing Unialgal Cultures

The algae loaded GF filter papers were immediately cultured using a method adapted from Ferris and Hirsch (1991) ('filter paper method'). Filters were placed on top of a stack of 5 Whatman GF papers saturated with a mixture of 1l BBM:BG media 1:1, inside a glass petridish with lid, before sealing with Parafilm. Samples were then left at room temperature and in low light (on lab bench) until colonies appeared.

Small samples were taken from colonies and placed on a microscope slide with a coverslip for quick examination under microscope Nikon Eclipse 90i and confocal D-Eclipse. Colonies with green or blue-green pigmentation (either single celled, filamentous or of diatom morphology) were picked off, diluted and re-plated on filter paper as described above, as well as on BBM pH 6.5 and BG-11 pH 7.1 agar plates (1% agar). Where possible, both filter paper and agar methods were repeated until a unialgal culture was derived from single colonies comprised of one alga (assessed under inverted microscope Nikon eclipse TE2000-S).

The method of (Vaara et al. 1979) was used ('agar plate scoring method') to clean filamentous samples of contaminant fungal species. Agar plates were scored perpendicular to the end of inoculation. Plates were covered in black cloth with the end opposite the inoculation site exposed to a light source. This arrangement encourages growth of algal filaments along the agar scores, towards the light, thereby shedding fungal and bacterial contaminants.

All isolates were deposited in the CCAP culture collection and have been assigned the following collection numbers. *Oscillatoria sancta* (CCAP 1459/46), *Microcoleus chthonoplastes* (CCAP 1449/2), *Mastigocladus laminosus* (CCAP 1447/3), *Klebsormidium sp.* (CCAP 335/20), *Coelastrella saipanensis* (CCAP 217/9), *Chroococcidiopsis thermalis* (CCAP 1423/1) and *Hantzschia sp.* (CCAP 1030/1).

Examining Temperature Tolerance

An experiment to determine a suitable temperature range for further analysis was carried out using heat blocks. All isolates together with a reference alga, *Chlorella emersonii* (*Ce*), were cultivated in triplicate in 15 ml Sterilin falcon tubes at temperatures between 25-60°C at intervals of 5°C. Tubes were shaken manually twice a day and the lids opened once a day to allow for gas exchange. The growth was assessed visually by comparing the tubes across the temperature ranges. As a result of the experiment 20°C, 30°C and 40°C were chosen for further experiments.

Each species was cultivated in 2×250ml conical flasks with 100 ml of suitable media: Bolds Basal Medium pH 6.5 (BBM) for eukaryotes and Blue Green-11 Medium pH 7.1 (BG-11) for prokaryotes and a 'Diatom Media' for the diatom (DM), comprised of BBM with the addition of 100µl B vitamins (0.0024% $B_1B_7B_{12}$ 1:1:1) and 100µl sodium silicate. The flasks were inoculated with 100µl stock cultures. Cultures were incubated in a plant growth chamber (in Sanyo MLR-351) on a rotary platform (Sanyo MIR-S100) at 100 rpm, with a 100-150 µmol m^{-2} s^{-1} of light on a 12 hour light/dark cycle. *Chroococcidiopsis thermalis* and *Hantzschia* sp. culture flasks were covered in a double layer of neutral density filter (StageElectrics, Bristol, UK) due to their sensitivity to light.

After 12 days one flask of each culture was pelleted and re-suspended in an appropriate nitrogen free media and cultivated for a further 6 days. Samples were then pelleted by centrifugation and lyophilised.

For single celled algae, growth measurements were recorded as cell counts using a haemocytometer and dry weights calculated by lyophilising 2 ml of the culture. Filamentous algae had a tendency to form mats or aggregates in the flasks and to stick to the walls of the vessel. Consequently, growth measurements for filamentous algae were done by initiating cultures in pre-weighed 15 ml falcon tubes containing 7.5 ml media. The algae were continuously mixed using a bloodtube rotator (Stuart Scientific SB1), Stone, Staffordshire, UK. At 3 day intervals 3 tubes were removed, centrifuged to remove water and lyophilised.

Identification of Isolates by DNA Barcoding

Algal cultures were grown in 100 ml of appropriate liquid media under culture conditions described above. DNA extracted from 50 ml of each culture was centrifuged and the wet pellet transferred to a 1.5 ml Eppendorf. 300 µl of 20 mM Na-phosphate buffer (pH 8), 150 µl of lysis solution (10% sodium dodecyl sulphate, 0.5M Tris HCl (pH 8),

0.1M NaCl) and sufficient 0.1 mm silica/zirconium beads added to make a paste which was then ground for 5mins on ice using a bench drill fitted with an Eppendorf micropestle. 300 µl DNA extraction buffer was added and centrifuged 13,000 rpm for 1 min; the supernatant was transferred to a fresh 1.5 ml Eppendorf. This process was then repeated. An equal volume of phenol-chloroform-isoamylalcohol (25:24:1) was added, mixed well and centrifuged for 3 min at 10,000 xg. The upper aqueous layer was then transferred to a fresh 1.5 ml Eppendorf. The phenol-chloroform-isoamylalcohol extraction repeated. DNA was precipitated with an equal volume of 70% ethanol and the pellet washed with 3 rounds of 70% ethanol followed by centrifugation (13,500 xg, 5 mins). Finally, the ethanol was removed and the pellet dried at room temperature for Y-X mins before resuspending in 50 µl of sterile deionised H_2O.

0.5 µl of DNA was added to a 200 µl thin walled PCRtube containing 12.5 µl DreamTaq Green Master Mix (Fermentas, UK), 1 µl forward primer and 1 µl reverse primers, 10 µl DNase-free water. Cycling conditions were 95°C for 5 min followed by 32 cycles (95°C 45s, X°C 45s, 72°C 120s (where X is the annealing temperature for each primer, indicated in with a final cycle of 72°C for 10 min. DNA sequences were checked manually, corrected and assembled using Sequencher 4.10.1 (Gene Codes Cooporation, Ann Arbor, Michigan, USA), sequences were assembled using corrected sequences were used to interrogate the NCBI online database using BLAST. The attribution of genus and/or species identity to the isolates was based on the BLAST total score. Total score is not included in the results (Table1) as it summarises and compares all data from resultant matches for a single query (arbitrary value for each sequence analysed). Instead, Table 1 contains 'query coverage' and 'max ident' which better describe the 'quality' of each final match. In order to be consistent 'total score' from the BLAST outputs was used as a means of identification. Where scores were ambiguous, decisions were based on morphology at the light-microscopy level.

Table 1: Identification of photosynthetic isolates using DNA barcoding of U16S/U18S rDNA gene

Isolate	Abbr.	Division	GenBank match ref.	Coverage%*	Max Ident%**	Max Score***	Other (NCBI no)
Coelastrella saipanensis*	Cs	Chlorophyta	AB055800	100	99	3090	BLAST and morphology (JX316760)
Klebsormidium sp.	K sp.	Chlorophyta	FR717537.1	98	99	1929	(JX316761)
Hantzschia sp.*	H sp.	Bacillariophyta					Morphology
Chroococcidiopsis thermalis	Ct	Cyanophyta	AB039005.1	100	99	2488	(JX316762)
Microcoleus chthonoplastes*	Mc	Cyanophyta	EF654089.1	96	91	1844	
Mastigocladus laminosus	Ml	Cyanophyta	AB607204.1	95	99	2385	(JX316764)
Oscillatoria sancta	Os	Cyanophyta	AF132933	96	99	1522	

Sequences compared to the NCBI online database using BLAST. Matches based on 'Max Score'. Date accessed 17.01.2012.

*% of sequence covered by database hit **% similarity of sequence covered by hit ***The score is calculated from the sum of the match rewards and the mismatch (independently for each segment).

Smith-Bädorf et al.

Smith-Bädorf et al. AMB Express 2013 3:9, doi:10.1186/2191-0855-3-9

H. sp. was identified based on morphology, using images taken from a slide preparation outlined below. An obtained pellet of diatom cells was washed with deionised water twice and centrifuged at 3000 rpm over 10 mins. Supernantant was removed and pellet resuspended in 1ml 30% (w/v) hydrogen peroxide and mixed well. Tube then placed in a water bath at 80-90°C for 1 hr and then allowed to cool. Sample was transferred to 15 ml falcon tube with ~10 ml deionised H_2O, mixed and centrifuged at 3000 rpm for 3 min. Supernatant then removed and pellet washed twice with distilled water, centrifuging each time and removing the supernatant. Working in a clean flow hood 0.5 ml of mixed suspension was dropped onto coverslips and dried using a hotplate at ~50°C. 1 drop of DPX mounting medium was dropped onto glass slides and using forceps the coverslips with dried diatom were inverted and placed over the drops of DPX. Assembled slides were then dried on the hotplate (at 80°C) overnight. Images were assessed by Dr. S. Spaulding (Colorado State University).

NCBI accession numbers for the Roman Bath isolates are as follows; *Coelastrella saipanensis*JX316760, *Klebsormidium* sp. JX316761, *Hantzschia* sp. JX316762, *Chroococcidiopsis thermalis*JX316763 and *Masticocladus laminosus* JX316764. For the purposes of this paper isolates *Os* and*Mc* have been identified as *Oscillatoria sancta* and *Microcoleus chthonoplastes*. However both sequences contained alignment gaps not acceptable to the NCBI database. Both isolates require re-sequencing for more accurate species identification.

Staining for Lipids

100 µl of cell culture was added to 25 µl DMSO and 1 µl 1% nile red dye in acetone in an eppendorf, mixed well and allowed to stain in the dark for 10 min. This was followed by dilution to 1 ml. Samples were viewed under microscope a Nikon Eclipse 90i and a confocal D-Eclipse C1. Nile red produces strong orange fluorescence at 543 nm/598 nm (excitation/emission) and BODIPY produces a strong green fluorescence at 493 nm/503 nm.

Oil Extraction and Transesterification

Total lipids were extracted from samples of algal biomass essentially as described by Bligh and Dyer (1959). Lyophilised algal samples were weighed and placed in MP Biomedical 'lysing matrix E' (Cambridge, CB1 1BH, UK) tubes together with 1.5 ml sterile deionised water. Samples were beaten in a Fastprep FP120 beadbeater (Thermo Scientific Savant, Surrey, UK) at setting 6.5 at intervals of 45 seconds for a total of 3 minutes. Tube contents were removed via a pipette and rinsed with a $CHCl_3$:MeOH (2:1) mixture. For each sample ~5 ml $CHCl_3$:MeOH (2:1) was added and the sample vortexed. Samples were then centrifuged at 3000 rpm for 5 min. The bottom layer was removed to a round bottom flask. Two subsequent rinses were performed to remove the remaining oils by adding 1.5 ml $CHCl_3$ to the sample, vortexing and centrifuging. Giving a total sample volume of 10 ml.

To transesterify the samples, MeOH (10 ml) and 18M H_2SO_4 (1 ml) were added to the sample, which was then held at reflux for 4 hours. The crude mixture was washed twice with H_2O (50 ml) to remove the acid and resulting glycerol. The organic layer was isolated, the solvent was removed under vacuum and the samples were dried and dissolved in 2 ml of 1,4-dioxane (99+%) and analysed by an Agilent 7890A Gas Chromatograph with capillary column (60m × 0.250mm internal diameter) coated with DB-23 ([50%-cyanpropyl]-methylpolysiloxane stationary phase (0.25m film thickness) and a He mobile phase (flow rate: 1.2ml/min) coupled with an Agilent 5975C inert MSD with Triple Axis Detector. The column was pre-heated to 150°C, the temperature held for 5 minutes and then heated to 250°C (rate of 4°C/min, then held for 2 min). The samples were quantified by comparison to known standards purchased from Sigma Aldrich.

RESULTS

Observations, Water Analysis and Identification of Isolated Photosynthetic Microorganisms

The Roman Baths appear to have extensive microbial mat communities submerged beneath the water surface, some with striking blue-green patches. There are also filamentous 'hairy' brown mats that grow around evolved gas bubbles. These eventually create gelatinous balloons-like structures that rise to the surface, anchored to the microbial mat below by filamentous 'ropes'. A bright orange muddy residue, possibly caused by bacteria, detritus and iron hydroxide precipitates form a thick layer (~2-3 cm), between the mats and stone surfaces. In addition to iron, the spring water contains high concentrations of calcium, sulphate, sodium and chloride (Kellaway, 1991). There are also visible mineral deposits, which form dependant on water agitation, cooling and oxygenation (Kellaway, 1991). Water analyses of the Great Bath (GB) and Kings Bath (KB) show that the abiotic conditions in the baths have remained stable since records began in 1874 The temperature of the Great Bath and Kings Bath were 39.0°C and 45.0°C respectively, measured 30cm below the surface.

In Figure 1, images of the isolates from the Roman Baths, alongside the reference alga *Chlorella emersonii* are shown. These are *Chlorella emersonii* (A), two green algae (B, C), a diatom (D) and four cyanobacteria (E, F, G, H). Isolates were identified as follows and will be referred to in figures in the same order by their abbreviations in brackets; B: *Coelastrella saipanensis* (Cs), C:*Klebsormidium* sp. (K sp.), D: *Hantzschia* sp. (H sp.), E: *Chroococcidiopsis thermalis* (Ct), F:*Microcoleus chthonoplastes* (Mc), G: *Mastigocladus laminosus* (Ml) and H: *Oscillatoria sancta* (Os). It should also be noted that there appears to be an abundance of microalgal species present in the baths, more than are discussed in this paper. Filamentous cyanobacterial species (*O.sancta, M.chthonoplastes, M.laminosus*) were isolated by from sample scrapings of microbial mats and contaminants removed using the 'agar plate scoring method'. All other species were isolated using the 'filter paper method'.

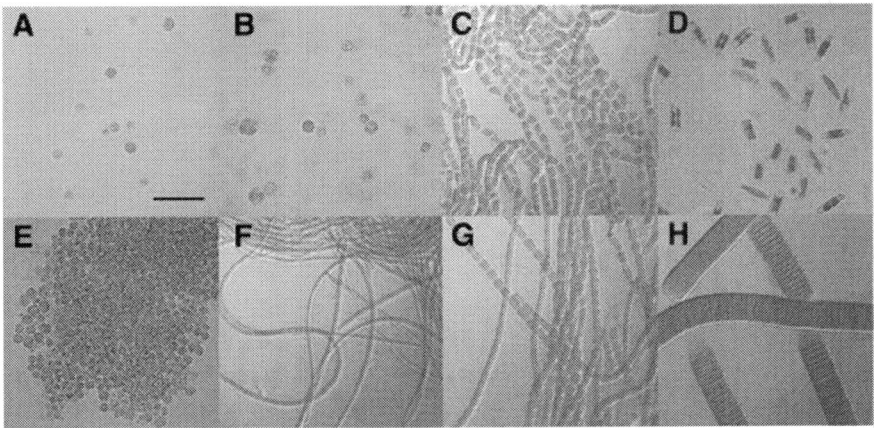

Figure 1: Light microscope images of algae isolated from the Roman Baths (x60). All algae were isolated from the Great Bath unless indicated (*=isolated from Kings Bath). A, *Chlorella emersonii* (*Ce*); B,*Coelastrella saipanensis* (*Cs*); C, *Klebsormidium* sp. (*K* sp.); D,*Hantzschia* sp. (*H* sp.); E, *Chroococcidiopsis thermalis* (*Ct*); F*, *Microcoleus chthonoplastes*(*Mc*); G*, *Mastigocladus laminosus* (*Ml*); H*, *Oscillatoria sancta* (*Os*). Bar=20 µm.

Microalgae are often 'elastic' in their ability to alter their morphology based on their growth conditions. In a unique environment such as the baths, algae may potentially look different to species standards. In addition, it is not uncommon for algae which are distantly related to look similar and vice versa. As such identifying algae based on morphology is very difficult, therefore DNA barcoding was the preferred method for identification of isolated photosynthetic microorganisms. Isolates were assigned species or genus level identification using Basic Local Alignment Search Tool (BLAST) using the 'total score' values and in some cases images from online culture collections to confirm matches (see methods). In order to be consistent 'Total score' from the BLAST outputs was used as a means of identification. Any isolates with multiple hits of the same score, visual identification was used to help confirm an ID.

The isolate identified in Table 1 as *C. saipanensis*, initially received identical BLAST scores for*Coelastrella saipanensis* and *Ettlia texensis*. Upon closer examination both entries in the database comprised of identical nucleotide sequences. Despite the difficulty in identifying algae based on morphology alone, images from online culture collections were used in order to identify this isolate from the two identical database 'hits'. Images of *Coelastrella* sp. in online culture collections showed similar cell granularity ('speckled' inclusions) to *C. saipanensis* but did not show aggregates of diving cells inside mother cells (a feature of *Ettelia* spp.). Therefore has been identified as *C. saipanensis*. For *K.* sp. there were many similarly scored species of the genera *Klebsormidium*. However, there were no distinct physical features present to identify a species based on morphology.

Identifying *H.* sp. to the species level proved problematic. There were many very similarly scored matches from different groups of diatoms that are morphologically very distinct. Diatoms, unlike many unicellular algal groups, can be more easily identified using morphology alone to the expert eye. Identification of *H.* sp. was subsequently based on images from a slide preparation (described in methodology), which were subsequently assessed by Dr. S. Spaulding (Colorado State University) for identification. For *M. chthonoplastes* most of the outputs in the BLAST search regardless of sorting had a ~98% query coverage from uncultured clones and *Microcoleus chthonoplastes* strains (with max idents of 91%). However, *Geitlerinema* cf. *acuminatum,* had 77% query coverage (percentage of the input sequence covered by individual sequences in the database) but a 97% max ident score (maximum percentage of high scoring pairs in a segment). Little has been published on *Geitlerinema* spp. but a few culture collection images do more closely match the growth morphologies of the isolate (*Mc*) than *Microcoleus* i.e. *M. chthonoplastes* isolate grows as an amorphous tangle of filaments like *Geitlerinema* as opposed to discrete bundles of fibres like *Microcoleus* spp.

Roman Bath isolate sequences for *Oscillatoria sancta* and *Microcoleus chthonoplastes* were not accepted to the NCBI database due to alignment gaps and some chimeric sequences. This may be attributed to their very thick gelatinous outer cell walls which likely contain some contaminant bacteria. Both isolates require re-sequencing for more accurate species identification.

Temperature Tolerance

Results of the temperature tolerance experiments show that the eukaryotic algae (*C.emersonii*, *C.saipanensis*, *Klebsormidium* sp. and

Hantzschia sp.) behave similarly and are able to grow at 30°C but do prefer lower temperatures. The cyanobacterial isolates have more variation between them but all show good growth at 35°C and some growth at 40°C, unlike the eukaryotes. Based on these results further growth experiments were carried out at 20°C, 30°C and 40°C, to give a low, medium and high value for comparison.

During isolation and culture maintenance of the Roman isolates it was noted that *O. sancta, C.thermalis* and *H.* sp. showed sensitivity to the light intensity used to culture stocks of *C.emersonii*(200 μMol m^{-2} s^{-1}) (data not shown). For this reason, it was decided to use a 'moderate' light intensity (80 μMol m^{-2} s^{-1}) and shaking speed for temperature tolerance experiments. Although this allows for results to be comparable for a specific set of culture conditions, the culture parameters are likely to have been suboptimal for a number of the species examined (e.g.*C.emersonii* potentially extending to the other green algae *K.sp, C.saipanensis* and even the cyanobacterial isolates). Growth media may also be suboptimal.

To further examine the growth rate under nitrogen enriched conditions the microalgae were cultured over 12 days and sampled periodically. To clearly communicate this information the final dry weights are presented in Figure 2, These experiments supported the preliminary data, where the eukaryotic algae all accumulated biomass at the lower temperature of 20°C, were more productive at 30°C but generally did not grow at 40°C. *K.* sp. and *C. saipanensis* accumulated the most biomass at 30°C, comparable to the reference alga *C. emersonii*. Within the cyanobacteria, the single celled species*C. thermalis* showed better growth at lower temperatures. *M. chthonoplastes, M. laminosus* and *O. sancta* acquired the most biomass at 30°C, with *M. chthonoplastes* and *M.laminosus* showing better growth at 40°C than 20°C. This was not found to be the case for *O.sancta* whose growth was impaired at 40°C. *M. laminosus* acquired the most biomass of all the cyanobacteria, at a temperature of 30°C.

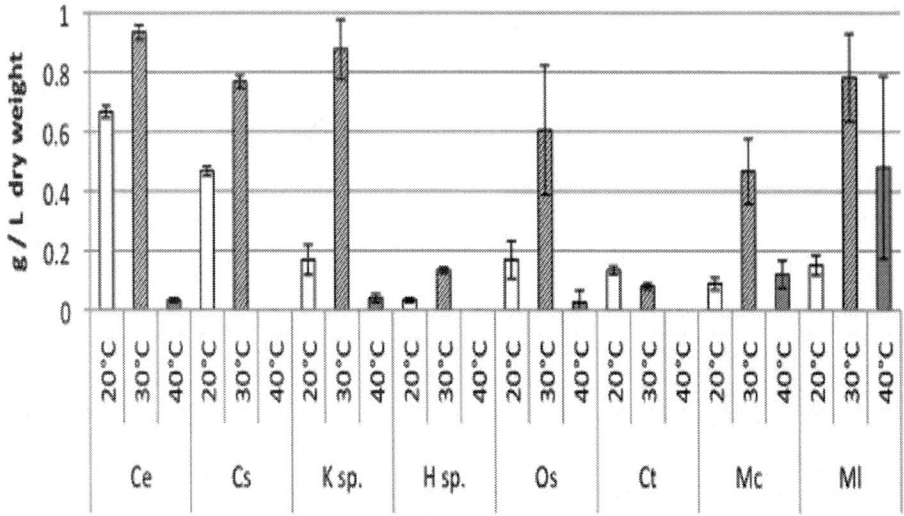

Figure 2: Growth of *Ce* and Roman Bath isolates in nitrogen sufficient medium after 12 days, measured as dry weight (g l-1).

Of all the algae tested, *C. emersonii* accumulated the most biomass at 20°C, *K* sp. the most at 30°C and *M.laminosus* the most biomass at 40°C. Both *H. sp.* and *C. thermalis* were not sufficiently productive at any of the temperatures investigated, achieving <0.2 g l^{-1} after 12 days, unlike other species which achieved between 0.5-0.9 g l^{-1} after 12 days.

Cytochemical Quantitation of Neutral Lipid Content

Staining gave a visual indication of the neutral lipid products contained in a sample (Figure 3). Nile red stains lipids and fluoresces red, where the lipids are predominantly unsaturated and yellow where the lipids are saturated. Photosynthetic pigments including chlorophyll present in eukaryotic algae and cyanobacteria auto-fluoresce at the same excitation wavelengths as nile red. This can make it difficult to identify stained lipids against the autofluorescent 'backdrop' of the cell. However the fluorescence of these pigments is useful in assessing the health of the cell as stressed cells undergo breakdown of these pigments. This can be seen in Figure 3A and B showing nitrogen starvation of *Klebsormidium* sp. at 20°C. At 20°C *K*. sp. (Figure 3A and B) has clearly visible lipid droplets and a noticeable lack of chlorophyll. At 30°C however, during nitrogen starvation, *K* sp. shows little or no breakdown in chlorophyll or accumulation of lipids (Figure 3C and D).

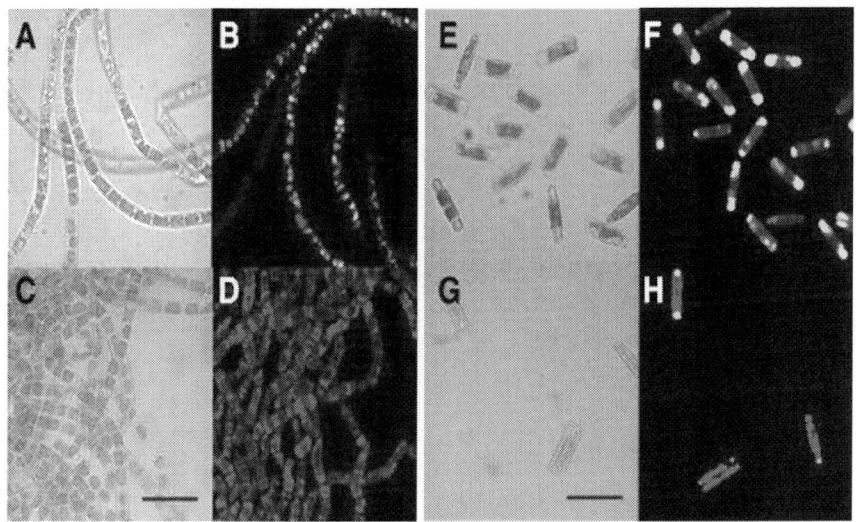

Figure 3: Microscope images of *Klebsormidium* sp. cultivated at 20°C (A,B) and 30°C (C,D) and starved of nitrogen. Microscope images of *Hantzschia* sp. cultivated at 30°C with sufficient nitrogen (E,F) and starved of nitrogen (G,H). Stained with nile red. Bar=20 μm.

Nitrogen starvation usually encourages accumulation of lipids in most microalgae (Cha et al. 2011). However, *H*. sp. at 30°C shows higher accumulation of lipids under nitrogen sufficient conditions (Figure 3E and F) compared to nitrogen starved conditions (Figure 3G and H).

Lipid Yield

To assess the effectiveness of the microalgae to produce biodiesel, the neutral lipids from the microbial samples (the weight of which is given in Figure 2) were isolated, using a chloroform and methanol solvent extraction and then converted into FAME through acid-catalysed transesterification and analysed by GC-MS.

Reducing the nitrogen content in the culture medium for green algae generally increases the neutral lipid accumulation in the cell (Cha et al. 2011). This effect is illustrated by the reference species in *Chlorella emersonii* where the FAME produced from the sample increased from 23 wt% to 34 wt% of the dry weight under nitrogen starved conditions at 20°C (Figure 4). A similar effect was seen with *K.* sp. (increasing from 1.9wt% to 16wt%) and *H.* sp. (from 31.9 to 40.5%). In contrast nitrogen-depletion did not change the amount of FAME recovered dramatically for *C. saipanensis*. Irrespective of any other effect, the green algae produced less neutral lipid at higher temperatures. Nitrogen-starvation of cyanobacteria had a detrimental effect on the lipid percentage at both 20°C and 30°C. At 40°C, however, a high percentage of FAME (10.4-45.6%) was recovered from *O. sancta* under most conditions (with the exception of 30°C). This effect was not observed in *M. laminosus*, *M. chthonoplastes* or *C. thermalis*.

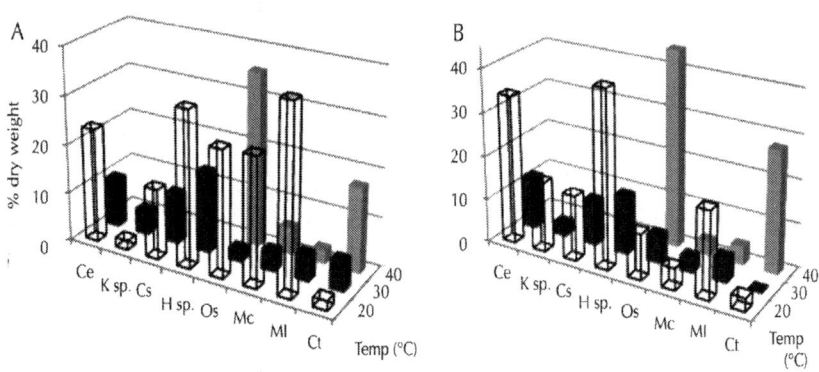

Figure 4: FAME produced from the Roman Bath isolates as a percentage of dry weight grown under A) nitrogen enriched and B) nitrogen starved conditions.

Reducing the nitrogen content in the growth media may have a detrimental effect on growth rate. Where this is the case, oleaginous microbes will fail to produce a substantial amount of biofuel on a per unit biomass basis (Figures 5 and 6). The highest biodiesel production was observed for *C. emersonii* and *C. saipanensis* at 20°C under nitrogen enriched conditions, where the lower content of glyceride lipids within the biomass were compensated by a higher growth rate. *K.* sp. and *H.* sp. biomass increased in the glyceride content when nitrogen content was reduced. None of the cyanobacteria produced a large amount of biodiesel irrespective of nitrogen conditions at 20°C and 30°C (Figure 5) compared to green algae *C. emersonii* and *C. saipanensis* (>0.15 g l⁻¹ of biodiesel). For the most part, this was due to low levels of biomass accumulation in the cyanobacterial isolates (Figure 2). However, at 40°C, nitrogen reduction resulted in recovery of a larger quantity of biodiesel from the cyanobacteria, though this represented only a small amount of biodiesel overall.

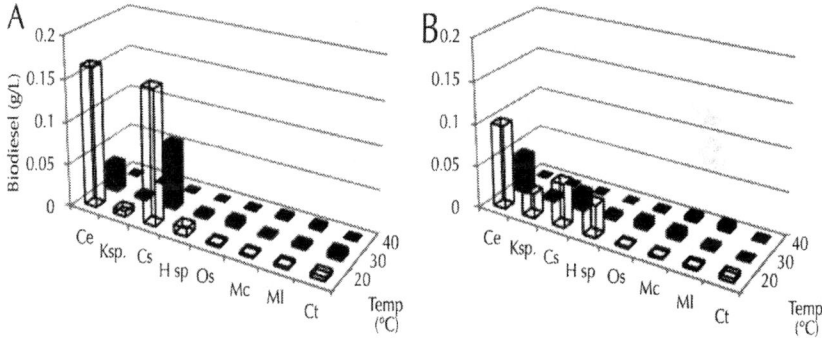

Figure 5: Total amount of biodiesel recovered from the Roman Bath isolates grown under A) nitrogen enriched conditions and B) nitrogen starved conditions.

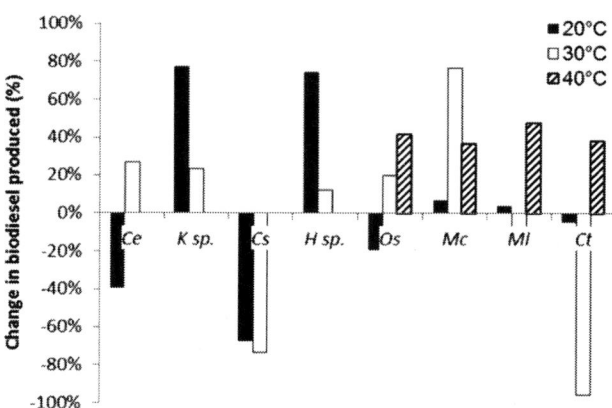

Figure 6: % change in the total amount of biodiesel produced on nitrogen starvation of the algal species.

Biodiesel Profile

The properties of biodiesel are highly reliant on the FAME profile (Fortman et al. 2008). Although chain length and the degree of unsaturation do have an effect on the fuel properties, the FAMEs can be placed into three loose categories to predict the fuel performance; saturated, monounsaturated and polyunsaturated esters (Figures 7 and 8). Biodiesel fuels rich in saturated esters have a superior cetane number but poor viscosity and low temperature properties; monounsaturated esters have acceptable low temperature properties and viscosity; fuels high in polyunsaturates have excellent low temperature properties and a low viscosity but a poorer cetane number and oxidative stability. Full FAME profiles were obtained for the isolated algae and are given in the On reduction of the nitrogen content during the culturing of *C. emersonii* the FAME profile did not change dramatically. However, an increase in temperature, in both nitrogen regimes, resulted in an increased polyunsaturated component predominantly at the cost of the monounsaturated fraction (Figure 7). A similar outcome was observed for *C. saipanensis*. In contrast to *C. emersonii*and *C. saipanensis* nitrogen starvation of *K.* sp. increased the production of unsaturated FAME whereas both the nitrogen enriched 20°C and 30°C samples contained predominantly saturated FAMEs. Like the green algae *C. saipanensis* and *C. emersonii*, the FAME profile of *H.* sp. did not change significantly following nitrogen starvation. However, raising the growth temperature (20°C – 30°C - 40°C) increased the proportion of the saturated component. The relatively high level of saturates recovered form both *K.* sp. and *H.* sp. would result in a poor quality biodiesel.

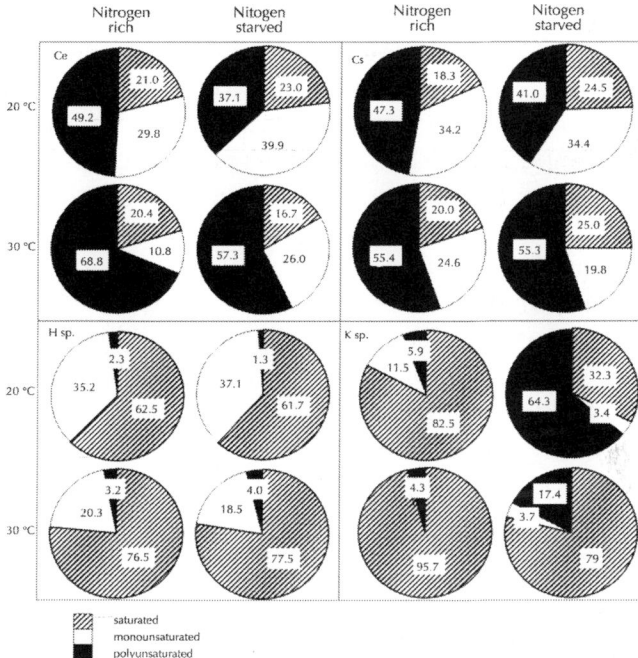

Figure 7: FAME profile of the microalgae cultured throughout this investigation.

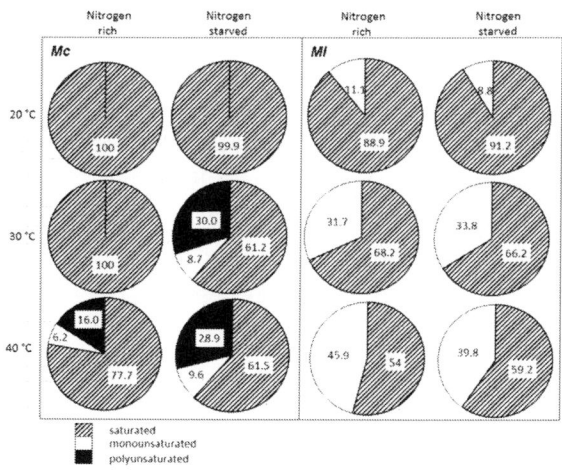

Figure 8: FAME profile of the cyanobacteria MC and ML.

The FAME profiles of the cyanobacteria were much simpler than those of the green algae. Unfortunately, insufficient biodiesel was recovered from *O. sancta* or *C. thermalis* to enable complete analysis of the FAME profile *M. chthonoplastes* produced only saturated esters at both low temperatures and in nitrogen-rich conditions at 30°C (Figure 8). Nitrogen depletion, or an increase in temperature to 40°C, resulted in an increase in unsaturated esters and composition more suitable for biofuel. *M. laminosus* produced no polyunsaturated esters under any of the conditions examined. At low temperatures there were around 90% saturates which decreased to approximately 66% when the temperature was increased to 30°C, and to approximately 50% at 40°C. Nitrogen-starvation had little effect on the FAME profile at any temperature.

DISCUSSION

Isolation and Identification of Microalgae

The excavation of the Roman Baths to uncover the present buildings and associated water-filled baths began in 1878 (Byrne 2008). Consequently, the opportunity for the colonisation of the baths by microalgae and other microbes has existed for some 134 years. Despite the city-centre location of the Baths and the relatively high water temperature initial sampling in the baths revealed the presence of multiple microalgal species. The isolates *O. sancta*, *M. chthonoplastes*, *M. laminosus* and *H. sp.*, proved the predominant species and were repeatedly found when attempting to isolate other species. Whilst the present work isolated and identified 7 species, our efforts were not exhaustive and further species remain for future isolation.

The first step to identifying algae present in the baths was to establish axenic unialgal cultures. Whilst other algae are seldom difficult to remove, bacteria are often physically associated with the muciferous layer of the algal cell walls and are not always eliminated by standard serial dilution and plating techniques. Impurities in agar (derived from algae) can also prove inhibitory to the growth of some algae and may encourage growth of bacterial and fungal contaminants. Some cyanobacterial isolates grew poorly on agar most likely due to toxic component in the agar (Dworkin and Falkow, 2006). Micromanipulation, filtration and use of chemicals have been successful with some groups (Ferris and Hirsch, 1991). The use of antibiotics appears attractive but must be used in a narrow concentration range to avoid damage to the chloroplast in eukaryotes and lethality to cyanobacteria (Issa, 1999).

In order to amplify and sequence 16S (cyanobacterial) or 18S (eukaryotic) rDNA gene sequencing, the isolates needed to be subjected to some decontamination. Filamentous cyanobacterial species (*O.sancta, M.chthonoplastes, M.laminosus*) were isolated by from sample scrapings of microbial mats and contaminants removed using the 'agar plate scoring method'. All other species were isolated using the 'filter paper method'. However, it is evident that the microbial mat communities present in the Roman Baths are required for microbial growth and survival. Hence it is also highly likely that resident algae have formed relationships with other organisms present (Croft et al.2006). Although most of the contaminants have been removed, the isolates are described as unialgal.

Employing a DNA-barcoding approach utilising the U16S and U18S rDNA gene sequences and a NCBI BLAST search as an identification tool proved problematical for *M.chthonoplastes,C.saipanensis, Hantzschia* sp. and *Klebsormidium* sp. This appears to arise due to both a relative paucity of gene sequences from microalgal species and the presence of incorrectly assigned sequences due to poor regulation of the NCBI database. Naturally, there is also a substantial bias toward more commonly used species. There is still uncertainty about the optimum fragments of DNA to use for identification (Surek, 2008). *Cox1* is the standard for most 'higher animals' but is not suitable for green algae and higher plants because its rate of evolution is too slow (Surek,2008). Common gene targets for algal identification are rRNA genes, mitochondrial genes and plastid genes, but each may fail to provide a conclusive result. Clearly there is a need to identify a single universal short DNA fragment that gives a clear identification of species (Surek, 2008).

Differences in geographical location can result in changes in target sequences *within* a species. This has been demonstrated for *M. chthonoplastes* (Garcia-Pichel et al. 1996) and *Synechococcus*sp. (Miller and Castenholz, 2000). Whilst this type of analysis has provided evidence that more thermotolerant lineages evolved from less thermotolerant ancestors (Miller and Castenholz, 2000), such ecologically induced changes can further complicate DNA sequence-based species identification.

Where DNA-barcoding proved inconclusive, algae were subjected to morphological examination at the light microscope level and images compared to online culture collections (e.g. the Culture Collection of Autotrophic Organisms (CCALA)). This approach proved successful for *C.saipanensis*but not for *H*.sp. Microalgae often lack distinct morphological features that can make them hard to identify (Pulz and Gross, 2004). Moreover, algal development can be plastic with cells changing shape and size during the lifecycle and in response to changes in culture conditions (Cheng et al.2011). Consequently, algal identification and taxonomy demands a 'polyphasic' approach that uses both molecular and morphological data as employed in this study (Surek, 2008).

Thermotolerance

Cyanobacteria are often found in warm, low nutrient environments (De Winder et al. 1990), and are typically more thermotolerant than eukaryotic algae (Barsanti and Gualtieri, 2006). Members of the single-celled cyanobacterial genus *Thermosynechococcus* are capable of surviving at 73-74°C, whilst thermophillic filamentous cyanobacteria typically occupy a lower temperature range of 55-62°C (Seckbach, 2007). In accord with this the present study found 50% of the isolates from the Roman Baths were cyanobacteria, and that these species remained viable and productive at higher temperatures than the eukayotic isolates.

Eukaryotic algae are generally absent from environments above 56-60°C largely due to the instability of organellar membranes at high temperatures (Tansey and Brock, 1972). Members of the Rhodophyta such as *C. caldarium* have been found to withstand 57°C (Seckbach, 2007). However, none of the Roman Bath isolates were from this genus. The comparatively poor growth of the eukaryotic isolates compared to the cyanobacteria, suggests that they were present in cooler locations within the baths i.e. on exposed surfaces close to the waterline.

In some cases there was a lack of correlation between growth rates at high temperatures *in vitro*and the prevalence of a particular species in the Baths. For example, *O. sancta* showed limited growth at 40°C (Figure 2) despite being the dominant microalgal species in the baths where it forms filamentous 'balloons' around evolved gas bubbles. In both environments the species is brown in colour rather than the typical blue-green. This is consistent with previous studies showing that high light or high temperature result in *Oscillatoria* sp. making a similar colour change and an associated reduction in growth rate (Gribovskaya et al. 2007). To ensure comparability of temperature effects all other conditions were *moderated* in order to be consistent for all species tested. For example, *C. emersonii* exhibits a faster growth rate at a higher light intensity over longer light: dark cycles, however during culturing these conditions were found to be detrimental to growth rate for *C. thermalis*, *H.* sp. and *O. sancta* (data not shown). Growth was therefore highly likely to be suboptimal for all the species investigated and further work is required to determine optimal growth conditions.

The Extreme Conditions in the Baths Favour Microbial Communities

The poor performance of unialgal cultures may reflect a requirement for cooperation in a microbial community. Microbial mats are important primary producers in extreme habitats where they provide environments that facilitate survival (Pattanaik et al. 2008). Evidence also suggests that cultured communities are more productive than monocultures (Croft et al. 2005). Microbial mats are prevalent in the Roman Baths and may therefore enable species to survive in this high temperature environment. The Roman Baths contain good levels of most trace elements required to support growth. However, some of these nutrients are in large excess which in turn can cause stress to algal cells. Calcium, sodium and chloride levels are much higher than standard media used for culturing these groups of algae (Barsanti and Gualtieri, 2006) and iron levels are much higher in the baths than most freshwater sources (Kellaway, 1991). In many aquatic environments iron is a limiting nutrient for productivity, yet in the baths it is at a similar concentration to most growth media. However it is important to note that for some species this could represent an excess. High iron concentrations have a negative effect on phytoplankton growth and increase oxidative stress (Estevez et al. 2001). Silicon is also plentiful in the bath, which presumably accounts for the abundance of *H.* sp.

Effect of Culture Conditions on FAME Production and Composition

Although nitrogen starvation and other stresses trigger neutral lipid accumulation in eukaryotic microalgae this is generally associated with a reduction in growth rate (Illman et al. 2000). This reduction was observed for *C. saipanensis* (Figure 2), where more FAME was produced under nitrogen starvation on a per cell basis (from 14.3 to 14.9 wt%) but had a low total productivity of biodiesel due to limited biomass production under these conditions (Figure 5). This trade off is therefore an important consideration when selecting microalgae for a particular production regime. Interestingly, although the cyanobacteria examined in this work accumulated a lower diversity of FAMEs than the eukaryotic algae, these species produced high levels of neutral lipids that were converted into FAME (up to 45 wt%) under nitrogen-rich conditions (Figure 5). The fact that cyanobacteria can accumulate lipids in the thylakoid membrane under conditions promoting photosynthesis and high growth rates may explain this behaviour (Karatay and Dönmez 2011).

The green algae *C. saipanensis* produced a large amount of polyunsaturated esters under all the conditions trialled similar only to the reference algae *C. emersonii* (refer to data). This FAME profile is roughly equivalent to sunflower or soybean oil (Knothe et al. 2005) and the resulting biodiesel would therefore have similar fuel properties including relatively high oxidative instability. Biodiesel produced from *H.* sp. and *K.* sp. was more saturated than that obtained from the other microalgal species (Figure 7). This effect increased with rising temperature (for *H.* sp. and *K.* sp. and increase in temperature from 20-30°C increased saturates by 14.0% and 13.2% respectively). Fuels high in saturates tend to have high cloud points, poor low temperature behaviour and a high viscosity (Knothe et al. 2005). Consequently, biodiesel produced from *H.* sp. and *K.* sp. would be unsuitable for high blend levels but would have combustion qualities akin to high performance diesel fuel such as a high cetane number, lower NO_x emissions and be highly oxidatively stable.

The cyanobacteria *M. chthonoplastes* and *M. laminosus* were rich in saturated esters (Refer to data). Although difficult to generalise, cyanobacteria tend to produce saturates in larger quantities than other species especially when cultivated at higher temperatures (Balogi et al. 2005; Nanjo et al. 2010; Dinamarca et al. 2011). Since unsaturated esters are more oxidatively stable this may be a direct response to a high temperature environment. Interestingly, the levels of saturated esters isolated from *M. laminosus* were reduced when cultured at higher temperature and at 40°C the biodiesel produced was almost 50% monounsaturated. Monounsaturated esters on balance have the most promising fuel properties, and biodiesel rich in these esters, such as rapeseed methyl ester or olive oil methyl ester, can be used at higher blend levels than other types of biodiesel so with careful control of the culture conditions biodiesel with suitable physical properties can be produced from the algal species isolated.

Our aim was to identify thermotolerant, oleaginous microalgae with potential as a source of renewable biodiesel. The Roman Baths proved to support a rich diversity of microalgal and cyanobacterial species. A total of 3 green algae, 1 diatom and 4 cyanobacteria were successfully established as unialgal cultures, and the majority assigned a species identification using a DNA barcoding approach. Whilst, a number of species produced high levels of neutral lipids, suitable for FAME production, under a range of conditions, all were more productive at lower temperatures than found in the Baths. Whilst these species do not sustain high productivity at extreme temperatures, an ability to survive temperature spikes in an open pond production system is an extremely desirable trait (Pulz and Gross, 2004). To date, the few species that are successfully cultivated commercially in open ponds are extremophiles able to grow in a highly selective environment (Xu et al. 2009). This work highlights the diversity in form, products and behaviour of algal species isolated from the same extreme environment and the importance of screening for new species. However a rapid species screening method requires a quick and efficient extraction method, which does not affect FAME profiles and ideally is scalable.

The culture conditions used to screen the microalgae were not optimised, suggesting that some of species could be developed into effective biodiesel producers. Four species, *K. sp., M. chthonoplastes, M. laminosus and O. sancta*, are filamentous, which could reduce harvesting and dewatering costs. One of these, *M. laminosus*, is also nitrogen-fixing species, which could again reduce input costs. Both these features could assist process on scale-up.

ACKNOWLEDGEMENTS

We would like to thank both Johnson Matthey and the DENSO Corporation for help both financially and materially, and Roger Whorrod for his generous endowment to the University resulting in the Whorrod Fellowship in Sustainable Chemical Technologies held by Dr. C. Chuck. We would like to thank Tom Byrnes and all the staff at the Roman Baths, Dr Miles Davis, Severn Trent Services for help with the water analysis and Dr. Sarah A. Spaulding of Colorado State University for help in identifying the diatom H. sp.

REFERENCES

1.   Andrews JN, Burgess WG, Edmunds WM, Kay RLF, Lee DJ (1982) The thermal springs of Bath. Nature 298:339-343

2.   Atkinson TC, Davison RM (2002) Is the water still hot? Sustainability and the thermal springs at Bath, England. Geological Society, London, Special Publications 198:15-40

3.   Balogi Z, Török Z, Balogh G, Jósvay K, Shigapova N, Vierling E, Vigh L, Horváth L (2005) Heat shock lipid in cyanobacteria during heat/light-acclimation. Arch Biochem Biophys 436:346-354

4.   Barsanti L, Gualtieri P (2006) Algae Anatomy Biochemistry and Biotechnology. CRC Press, Raton.

5.   Bligh EG, Dyer WJ (1959) A rapid method for total lipid extraction and purification. Can J Biochem Physiol 37:911-917

6.   Byrne T (2008) pers.comm. The Roman Baths, Abbey Church Yard, Bath, BA1 1LZ, UK.

7.   Cha TS, Chen JW, Goh EG, Aziz A, Loh SH (2011) Differential regulation of fatty acid biosynthesis in two Chlorella species in response to nitrate treatments and the potential of binary blending microalgae oils for biodiesel application. Bioresource Technol 102:10633-10640

8. Cheng Y-S, Zheng Y, Labavitch JM, VanderGheynst JS (2011) The impact of cell wall carbohydrate composition on the Chitosan flocculation of Chlorella. Process Biochem 46:1927-1933

9. Chisti Y (2007) Biodiesel from microalgae. Biotech Adv 25:294-306

10. Chisti Y (2008) Biodiesel from microalgae beats bioethanol. Trends Biotechnol 26:126-131

11. Chisti Y, Yan J (2011) Energy from algae: current status and future trends. Appl Energ 88:3277-3279

12. Croft MT, Lawrence AD, Raux-Deery E, Warren MJ, Smith AG (2005) Algae aquire vitamin B12 through a symbiotic relationship with bacteria. Nature 438:90-93

13. Croft MT, Warren MJ, Smith AG (2006) Algae need their vitamins. Eukaryot Cell 5:1175-1183

14. Cuvelier ML, Ortiz A, Kim E, Moeling H, Richardson DE, Heidelberg JF, Archibald JM, Worden AZ (2008) Widespread distribution of a unique marine protistan lineage. Environ Microbiol 10:1621-1634

15. De Winder B, Stal LJ, Mur LR (1990) Crinalium epipsammum sp. Nov. a filamentous cyanobacterium with trichomes composed of elliptical cells and containing poly-β-(1,4) glucan (cellulose). J Gen Microbiol 136:1645-1653

16. Dinamarca J, Shlyk-Kerner O, Kaftan D, Goldberg E, Dulebo A, Gidekel M, Gutierrez A, Scherz A (2011) Double mutation in photosystem II reaction centers and elevated CO_2 grant thermotolerance to mesophilic cyanobacterium. PLoS One 6(12):e28389

17. Dworkin M, Falkow S (2006) The Prokaryotes. Springer, Singapore.

18. Estevez MS, Malanga G, Puntarulo S (2001) Iron-dependant oxidative stress in Chlorella vulgaris. Plant Sci 161:9-17

19. Ferris MJ, Hirsch CF (1991) Method for isolation and purification of cyanobacteria. Appl Environ Microbiol 57:1448-1452

20. Fortman JL, Chabra S, Mukhopadhyay A, Chou H, Lee TS, Steen E, Keasling JD (2008) Biofuel alternatives to ethanol: pumping the microbial well. Trends Biotechnol 26:375-342

21. Garcia-Pichel F, Prufert-Bebout L, Muyzer G (1996) Phenotypic and phylogenetic analyses show Microcoleus chthonoplastes

to be a cosmopolitan cyanobacterium. Appl Environ Microbiol 62:3284-3291

22. Gressel J (2008) Transgenics are imperative for biofuel crops. Plant Sci 174:246-263

23. Gribovskaya IV, Kalacheva GS, Bayanova YI, Kolmakova AA (2007) Physiology-biochemical properties of the cyanobacterium Oscillatoria deflexa. Appl Biochem Microbiol 45:285-290

24. Groom MJ, Gray EM, Townsend PA (2008) Biofuels and biodiversity: principles for creating better policies for biofuel production. Conserv Biol 22:602-609

25. Haik Y, Selim MYE, Abdulrehman T (2011) Combustion of algae oil methyl ester in an indirect injection diesel engine. Energy 36:1827-1835

26. Hu Q, Sommerfeld M, Jarvis E, Ghirardi M, Posewitz M, Seibert M, Darzins A (2008) Microalgal triacylglycerols as feedstocks for biofuel production: perspectives and advances. Plant J 54:621-639

27. Illman AM, Scragg AH, Shales SW (2000) Increase in Chlorella strains calorific values when grown in low nitrogen medium. Enzyme Microb Tech 27:631-635

28. Issa AA (1999) Antibiotic production by the cyanobacteria Oscillatoria angustissima and Calothrix Parietina. Environ Toxicol Pharmacol 8:33-37

29. Karatay SE, Dönmez G (2011) Microbial oil production from thermophile cyanobacteria for biodiesel production. Appl Energ 88:3632-3635

30. Kellaway GA (1991) Hot Springs of Bath: Investigations of the thermal waters of the avon valley. Bath City Council, Oxford.

31. Knothe G, Van Gerpen J, Krahl J (2005) The Biodiesel Handbook. AOCS Press, Campaign, IL.

32. Miller SR, Castenholz RW (2000) Evolution of thermotolerance in hot spring cyanobacteria of the genus Synechococcus. Appl Environ Microbiol 66:4222-4229

33. Mohan N, Hanumantha RP, Ranjinth KR, Sivasubramanian V (2010) Mass cultivation of Chroococcus turgidus and Oscillatoria sp. and effective harvesting of biomass by low-cost methods.

Available from Nature Proceedings http://dx.doi.org/10.1038/npre.2010.4331.1

34. Mutanda T, Ramesh D, Karthikeyan S, Kumari S, Anandraj A, Bux F (2011) Bioprospecting for hyper-lipid producing microalgal strains for sustainable biofuel production. Bioresource Technol 102:57-70

35. Nanjo Y, Mizusawa N, Wada H, Slabas AR, Hayashi H, Nishiyama Y (2010) Synthesis of fatty acids de novo is required for photosynthetic acclimation of Synechocystis sp PCC 6803 to high temperature. BBA- Bioenergetics 1797:1483-1490

36. Ono E, Cuello JL (2006) Feasability assessment of microalgal carbon dioxide sequestration technology with photobioreactor and solar collector. Biosystems Eng 95:597-606

37. Pattanaik B, Roleda MY, Schumann R, Karsten U (2008) Isolate-specific effects of ultraviolet radiation on photosynthesis, growth and mycosporine-like amino acids in the microbial mat-forming cyanobacterium Microcoleus cthonoplastes. Planta 227:907-916

38. Pittman JK, Dean AP, Osundeko O (2011) The potential of sustainable algal biofuel production using waste water resources. Bioresource Technol 102:17-25

39. Pulz O, Gross W (2004) Valuable products from biotechnology of microalgae. Appl Microbiol Biotechnol 65:635-648

40. Purton S, Stevens DR (1997) Review: Genetic engineering of eukaryotic algae: progress and prospects. J Phycol 33:713-722

41. Raja R, Hemaiswarya S, Kumar NA, Sridhar S, Rengasamy R (2008) A perspective on the biotechnological potential of microalgae. Crit Rev Microbiol 34:77-88

42. Ratha SK, Prasanna R (2012) Bioprospecting microalgae as potential sources of 'green energy', challenges and perspectives, Appl. Biochem Microbiol 48:109-125

43. Rittmann BE (2008) Opportunities for renewable bioenergy using microorganisms. Biotech Bioeng 100:203-212

44. Rupprecht J (2009) From systems biology to fuel - Chlamydomonas reinhardtii as a model for a systems biology approach to improve biohydrogen production. J Biotechnol 142:10-20

45. Saraf S, Thomas B (2007) Influence of feedstock and process chemistry on biodiesel quality. Process Safety and Environmental Protection 85:360-364

46. Schenk PM, Thomas-Hall SR, Stephens E, Marx UC, Mussgnug JH, Posten C, Kruse O, Hankamer B (2008) Second generation biofuels: high-efficiency microalgae for biodiesel production. Bioenergy Res 1:20-43

47. Seckbach J (2007) Algae and cyanobacteria in extreme environments. Springer, Dordrecht.

48. Song D, Fu J, Shi D (2008) Exploitation of oil-bearing microalgae for biodiesel. Chinese Journal of Biotechnology 24:341-348

49. Surek B (2008) Meeting report: algal culture collections, an international meeting at the culture collection of algae and protozoa (CCAP), dunstaffnage marine laboratory, Dunbeg, Oban, United Kingdom, June 8-11, 2008. Protist 159:509-517

50. Tansey MR, Brock TD (1972) The upper temperature limit for eukaryotic organisms. Proc Nat Acad Sci USA 69:2426-2428

51. Taton A, Grubisic S, Brambilla E, De Wit R, Wilmotte A (2003) Cyanobacterial diversity in natural and artificial microbial mats of lake Fryxell (McMurdo dry valleys, Antarctica): a morphological and molecular approach. Appl Environ Microbiol 69:5157-5169

52. Uduman N, Qi Y, Danquah MK, Forde GM, Hoadley A (2010) Dewatering of microalgal cultures: a major bottleneck to algae-based biofuels. J Ren Sus Energy 2:1-15

53. Um BH, Kim YS (2009) Review: A chance for Korea to advance algal-biodiesel technology. J Ind Eng Chem 15:1-7

54. Vaara T, Vaara M, Nieml S (1979) Two improved methods of obtaining axenic cultures of cyanobacteria. Appl Environ Microbiol 38:1011-1014

55. Varfolomeev SD, Wasserman LA (2011) Microalgae as source of biofuel, food, fodder and medicines. Appl Biochem Microbiol 47:789-807

56. Xu L, Weathers PJ, Xiong XR, Liu CZ (2009) Review: Microalgal bioreactors: challenges and opportunities. Eng Life Sci 9:178-189

Experimental and Theoretical Simulation of Sublimating Dusty Water Ice with Implications for D/H Ratios of Water Ice on Comets and Mars

John E Moores[1,2], Robert H Brown[1],
Dante S Lauretta[1], and Peter H Smith[1]

[1]Lunar and Planetary Laboratory, Department of Planetary Sciences, University of Arizona, 1629 E University Blvd, Tucson, AZ 85721-0092, USA

[2]Now at: Centre for Planetary Science and Exploration, Department of Physics and Astronomy, University of Western Ontario, 1151 Richmond Street, London, ON N6A 3 K7, Canada

ABSTRACT

Sublimation experiments have been carried out to determine the effect of the mineral dust content of porous ices on the isotopic composition of the sublimate gas over medium (days to weeks) timescales. Whenever mineral dust of any kind was present, the D/H ratio of the sublimated gas was seen to decrease with time from the bulk ratio. Fractionations of up to 2.5 were observed for dust mixing ratios of 9 wt% and higher of JSC MARS-1 regolith simulant 1-10 µm crushed and sieved fraction. These favored the presence of the light isotope, H_2O, in the gas phase. The more dust was added to the mixture, the more pronounced was this effect. Theoretical modeling of gas migration within the porous samples and adsorption on the excavated dust grains was undertaken to explain the results. Adsorption onto the dust grains is able to explain the low D/H ratios in the sublimate gas if adsorption favors retention of HDO over H_2O. This leads to significant isotopic enrichment of HDO on the dust over time and depletion in the amount of HDO escaping the system as sublimate gas. This effect is significant for planetary bodies on which water moves mainly through the gas phase and a significant surface reservoir of dust may be found, such as on Comets and Mars. For each of these, inferences about the bulk water D/H ratio as inferred from gas phase measurements needs to be reassessed in light of the volatile cycling history of each body.

BACKGROUND

Motivations for Simulating HDO

The deuterium to hydrogen ratio (D/H) of water is often used to infer the evolution of water inventories on planetary bodies [for instance, [1, 2]]. It is a particularly effective tool for several reasons. First, the signatures of D and H-bearing molecules can be detected and distinguished spectroscopically [3, 4] in the atmospheres of other planets. Secondly, there is a high potential for large mass-dependant fractionations upon dissociation as D and H have the largest mass difference of any isotope pair, with D being twice as massive as H. This must be considered along with the tendency of terrestrial planetary atmospheres to evolve towards more oxidated states as hydrogen is lost to space through Jeans escape from the atmospheric exobase [2]. If the water inventory of the planet is exchangeable with the atmosphere, the signature of this missing hydrogen will be partly preserved in the remaining water reservoir as a large enrichment in D/H.

This argument is often used to assert that planetary bodies with elevated D/H ratios in their present day-atmospheres lost most of their water during the early evolution of their atmospheres [5]. However, there are two important caveats to this conclusion. First, in order to assess the amount of fractionation it is necessary to make estimates of the initial D/H values of the reservoir in question[1]. This is a difficult task even for the bulk solar system as the initial store of deuterium in the solar system's largest reservoir, the Sun, was consumed early in the life of the solar system. The protosolar value can be obtained through analysis of the Jovian atmosphere and solar wind implanted isotopes [6], however, this remains more difficult for specific bodies. For instance, while [7] argues for a primitive origin for comets, two distinct reservoirs with different D/H ratios are required to explain the observed values.

The second caveat is that fractionation of the major planetary water reservoirs is not ongoing in the present era, i.e. since the surface and climate evolved to its present state. As a result, explanations for large variations in the observed D/H ratios between different Solar System bodies, such as the 5.5-fold enrichment of the heavy isotope seen in the Martian atmosphere [3,4] compared to VSMOW (Vienna Standard Mean Ocean Water) or the 15-fold enrichments in comets [8-10] compared to protosolar values [6,10] have largely ignored whether the enhanced signal observed is the result of a bulk loss of water, or simply a byproduct of cycling at cold temperatures in the presence of large quantities of dust. Given the fractionations that are possible on the Earth during volatile cycling between reservoirs [11], it is important to quantify this effect.

The isotopic fractionation of water ice during sublimation on planetary bodies has been considered a simple temperature-driven process in which colder ices tended to fractionate more heavily than warmer ones due to the differences in lattice binding energy between H_2O and HDO [12-16]. This energy difference results in lower equilibrium sublimation pressures for HDO, the more refractory form of water, and intensifies at lower temperatures, increasing the relative refractory nature of HDO compared to H_2O. Typically, when this information is applied to planet-scale analyses, either just this equilibrium value has been used [17] or is first combined with the concept of a static lattice in which no fractionation occurs due to a buildup of the more refractory isotope on the surface until the ice sublimates at the bulk ratio [18].

Recent Work on the Simulation of Sublimation and Fractionation of Water Ice

Additionally, recent laboratory experiments have shown that the sublimation of water ice is not a simple and steady process, and that D/H ratios in the vapor can be time variable, mimicking to a degree the Rayleigh-like fractionation seen in the evaporation of liquids [19]. Previous experimental work [20] provides a good summary also indicates that realistic samples containing mixtures of ices or contaminating dusts can severely impact the sublimation rate and isotopic character of the escaping gas. The observations of the KOSI (Kometen Simulation) experiments [21-28], designed to simulate sublimation on cometary bodies, are particularly relevant to the work to be described in this paper.

The KOSI experiments were a major initiative to understand the processes that affect comets and are made up of eleven individual experiments conducted at the DLR Institute of Space Simulation, Cologne. Collectively, these experiments examined many different compositions of comets and ice/dust ratios [20]. However, there are two areas where these experiments need to be extended in order to examine the long-term isotopic behavior of samples. First, none of the KOSI experiments lasted longer than 59 hours. Secondly, only two of the experiments, KOSI-7 and KOSI-11, used isotopic tracers and only KOSI-11 employed any HDO enrichment [20]. As such, even though there is data on the isotopic character of some of the remaining sample condensate [29] there are no D/H profiles of the sublimate gas available to identify how the sample changes with time.

This paper will fill this gap by considering the isotopic medium-term evolution of dusty ice. The results of an experimental investigation into the sublimation of realistic samples of porous, disaggregated and dusty ices will be described in order to study the evolution of fractionation between the solid and gas with time. Additionally, some theory about how realistic samples of ice will fractionate in planetary settings will be discussed as well as the implications that the results have for the evolution of Comets and the polar caps of Mars. Definitions of all variables used in this analysis are presented in Table 1.

Table 1: Summary of symbols used in equations

Symbol	Units	Definition
A	molec s kg^{-1}	Constant from Equation (1.a)
A_{surf}	m^2	Surface area of the frozen sample
A_s	m^{2}kg-1	Specific surface area of the regolith available for adsorption
α	Pa^{-1}	Langmuir adsorption constant (subscripts refer to the species)
B	m^{3}molec^{-1}	Constant from Equation (1.b)
C	kg m molec^{-1}s^{-1}	Constant from Equation (1.c)
χ	dimensionless	Mass fraction (subscripts: Z = of dust; H2O = of water)
D_{eff}	m^{2}s^{-1}	Diffusivity (subscripts: eff = Effective Diffusivity; 0 = time independent) When bolded represents a 2- element vector with [H_2O HDO]
e_{sys}	dimensionless	Pumping efficiency (pressure ratio between chamber and QMS-inlet)
G	dimensionless	Rate coefficient matrix
γ_Z	dimensionless	Volume fraction of dust in lag mantle
k	dimensionless	Rate constant, Superscripts: ads = adsorption, des = desorption, X = cross-reaction substitution by vapor-phase species. When bolded represents a 2- element vector with [H_2O HDO]
k_B	m^{2}kg s^{-2}K^{-1}	Boltzmann Constant (1.38 × 10^{-23})
J	molec m^{-2}s^{-1}	Molecular flux at a point in the apparatus
k	s^{-1}	Langmuir adsorption kinetic constant, Subscripts: H2O, HDO, X = substitution; Superscripts: ads = adsorption, des = desorption

l	m	Thickness of an adsorbed monolayer of water (2.75×10^{-10})
m_{H2O}	kg molec^{-1}	Mass of a single water molecule (2.99×10^{-26})
M_{H2O}	kg mol^{-1}	Molar Mass (0.01802 for water)
N_A	molec mol^{-1}	Avogadro's Number (6.02×10^{23})
n	molec m^{-3}	Number density of molecules
p	Pa	Pressure (subscripts: sat = Saturated Vapor Pressure)
R	m^{2}K^{-1}s^{-2}	Specific Gas Constant (462 for water)
R_u	J mol^{-1}K^{-1}	Universal Gas Constant (8.3145)
	kg m^{-3}	Density (subscripts: Z = of dust; reg = of regolith; H2O = of water)
	J m^{-3}	Thermodynamic Adsorption Constant
T	K	Temperature (subscripts: QMS = the tubing at the throat of the QMS)
t	s	Time (subscripts: 0 = initial time)
	dimensionless	Fraction of adsorption sites occupied by water molecules when bolded indicates a two-element vector with [H_2O HDO]
V	m^{3}s^{-1}	Pumping speed
z	m	Thickness of dust lag (subscripts: 0 = initial time)

Moores *et al. Planetary Science* 2012 1:2, doi:10.1186/2191-2521-1-2

METHODS

Experimental Apparatus

The experimental apparatus is detailed in [19]. This setup, shown schematically in Figure 1, consists of a vacuum-jacketed cryostat in which samples can be cooled from below by a closed-cycle helium cooler attached to a 1/2-inch copper thermal capacitor and heated from above by a Quartz-Tungsten-Halogen (QTH) lamp which produces a blackbody spectrum and was run at powers between 1 and

4 W. The chamber has a volume of 150 cm³ with a cross-sectional area of 10 cm² and typically contained 50 cm³ of sample. 3 silicon diode temperature sensors were located on the outside of the thin-walled chamber wall within the vacuum jacket. A high vacuum was maintained within the inner sample chamber with the turbomolecular pump of a Quadrupole Mass Spectrometer (QMS-100) atmospheric sampler from Stanford Research Systems. This instrument had a capacity of 20 liters per second at operational pressures, as verified by a calibrated helium leak. For initial evacuation a second Pfeiffer turbomolecular pump was operated on a separate line. This allowed the main analysis line to remain as pristine as possible. Several pressure measurement devices were located on this line including a 20mtorr capacitance manometer near the chamber. At the QMS-100 inlet, a cold cathode gauge verified the pressure measured just inside the QMS-100 using a Stanford Research Systems RGA-100 residual gas analyzer quadrupole mass spectrometer. By monitoring the partial pressures of masses 17-22 (primarily $H^{16}O$ through $D_2^{18}O$) the D/H ratio of the gas escaping the sample chamber could be measured.

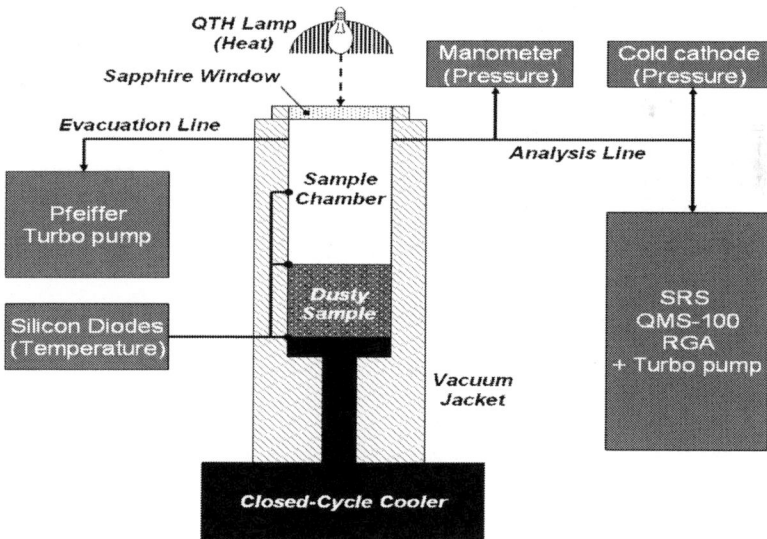

Figure 1: Schematic of Experimental apparatus showing major gas plumbing and sensors. The volume of the sample chamber is approximately 150 cm³.

Experimental procedure

A typical run consisted of three stages: sample preparation, sample insertion and sample observation. Sample preparation involved mixing measured volumes of H_2O and D_2O together with a measured mass of dry mineral dust samples (hereafter described as "dust"). Typically, 50 mL samples were prepared by combining 47.5 mL of H_2O with 2.5 mL of D_2O yielding an initial D/H ratio of 0.053 with the deuterated species present almost entirely as HDO due to rapid proton transfer. This liquid was then combined directly with an appropriate mass of dry dust to achieve the desired mass fraction, shaking well to incorporate all the components thoroughly. The properties of the dust used will be described in the "Dust and regolith simulants" section.

Sample insertion procedures evolved over the course of the experimental runs; however, differences in preparation do not appear to have had a noticeable effect on the results based on blank samples (see Section "Dust and regolith simulants"). Early on, the sample chamber was filled with liquid nitrogen (LN2) and the liquid water-dust mixture created during sample preparation was poured directly into the chamber in small amounts, shaking the remaining liquid between pours. Later on, a step was added in which small amounts of the prepared liquid were poured into a mortar filled with LN2 where the resulting dusty ice was crushed to sand-sized particles using a pestle. This water-dust-LN2 slurry was then transferred to the waiting LN2-filled sample chamber. The advantage of the additional step was that it produced a more even surface layer as the sand sized particles settled in the LN2.

LN2 was used as a medium for two reasons. First it was able to rapidly freeze and cool the ice-dust liquid mixtures before growing ice crystals were able to exclude the dust particles. Microscopy of prepared ices (Figure 2) shows that the dust is well incorporated into the final icy samples created using this method. Several different ice crystal morphologies were observed (e.g. hexagonal, banded, rounded grains) depending on the dust concentration, however the rounded textures shown in Figure2 were the most common. Secondly, LN2 boil-off during sample freezing prevented room air from condensing on the samples, hence preventing contamination by isotopically light water.

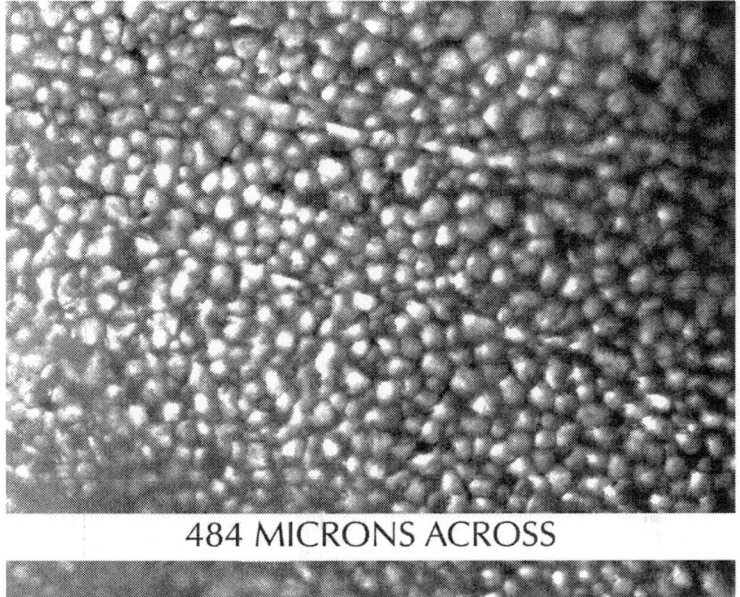

484 MICRONS ACROSS

Figure 2: Microscopy of 25 wt% 1.5 μm diameter TiO_2 dust in water ice following flash freezing in LN2. The ice grains formed are ~20 μm across and are separated by a matrix of the finer dust. Generally, lower concentrations of dust produced larger characteristic grain sizes.

Once the samples were inserted in the sample chamber, but before the LN2 had evaporated, a relief valve on the sample chamber was closed and the sapphire observation window placed upon its gasket. This allowed the sapphire to act as a pressure relief on the sample chamber without admitting room air until the nitrogen had evaporated. Once the nitrogen pressure had decreased to almost atmospheric pressure, the window was clamped down onto its gasket and a blow-down line was opened. This procedure cleaned the sample and removed any residual nitrogen. To remove any solid nitrogen created by this evacuation, the temperature of the base of the sample (hereafter referred to as «basal» temperature) was temporarily increased to 150 K. Note that the basal temperature is simply an experimental parameter. It is not the bulk temperature of the sample and not representative of the temperature of the actively sublimating surface. The actual temperature of the surface is most directly determined by the initial gas pressure and related to the basal temperature only by the thermal conductivity of the sample.

While the sample was being cleaned by the blow-down line, a background mass spectra was obtained on the valved-off analysis line by averaging 30 minutes worth of observations. This procedure was repeated after the experiment was concluded. Both observations were combined to provide an average background to be removed from the observed values of pressure for each species during the experimental run. Typical background values for the water group, masses 18 to 22, were between 1×10^{-9} torr and 2×10^{-8} torr with the light isotope $H_2^{16}O$, mass-18, being the largest component.

Finally, once a sufficiently low pressure had been achieved in the chamber (typically a few days), the sample observation phase could begin. At this point, the blow-down line would be shut, the analysis line opened up, and the lamp moved into position and turned on. Logging of all sensor outputs including all three temperature sensors, the basal temperature, the 20mTorr and cold cathode pressure gauges as well as the partial pressures of all species from the QMS would also be started. Generally, no external changes were permitted over the course of an experimental run.

During this stage, [19] operated with temperatures at the base of the sample near 150 K when dealing with solid ice samples to produce surface temperatures of 200-210 K. However, since more porous samples were employed here, lower basal temperatures were required, of order 60-120 K in order to produce the same surface temperatures and to maintain the QMS in its preferred linear range at or below 5×10^{-5} torr at the QMS inlet.

Dust and Regolith Simulants

Not including the blanks which contained no dust, five separate compositions and sizes of dusts and regolith simulants were used. These were: (a) Titanium (IV) Oxide (TiO_2 Rutile) in 1-2 μm sized particles (b) Silicon Dioxide in 1-2 μm sized particles (c) cuttings from the Fukang Pallasite in 1-50 μm sized particles (d) dry sieved JSC Mars-1 Regolith Simulant 76-105 μm sized particles and (e) crushed and water settled JSC Mars-1 Regolith Simulant 1-10 μm sized particles. Simulants (a) through (c) were used to achieve a broad understanding of the apparatus, to outline the effect of dust on samples and to perform blank runs without dust. These samples were chosen mainly for their availability and not as analogues for realistic planetary dust analogues. However, this paper will focus mainly on results from the last two simulants since these represent realistic planetary analogues for the Martian polar caps. It should be noted, however, that all dust stimulants exhibited similar trends (see Section «Comparison to other dust compositions and blank runs»).

JSC Mars-1 is a standard martian regolith simulant noted as spectrally and chemically similar to observations on the martian surface and consists of water-modified glassy volcanic ash collected from the Puʻu Nene cinder cone on the island of Hawaii [30]. The material is primarily SiO_2 (43.7 wt%) Al_2O_3 (23.4 wt%) and Fe_2O_3 (11.8 wt%) with smaller amounts of CaO (6.2 wt%), TiO_2 (3.8 wt%), FeO (3.5 wt%) and MgO (3.4 wt%) in addition to trace constituents.

Production of dust (d) from this material was straightforward using a sieving machine. However, in order to obtain dust (e), a grain size representative of martian airfall dust, it was first necessary to process the sieved JSC Mars-1. The grain size of martian airfall dust comes from the 1.6 μm radius particle peak of the distribution of dust observed in the martian atmosphere from landed spacecraft, as determined by studies of the martian atmosphere [31-33]. This material was produced by taking raw JSC Mars-1 and crushing it using a mortar and pestle before pouring the resulting detritus into a 10 cm deep water settling tank for 30 minutes. Those particles still in suspension were poured off along with the water and were desiccated and collected as chips. Microscopy confirmed the expected grain size distribution of particles up to 10 μm in diameter. However, the composition of this material was not assayed following collection and may be slightly different from the proportions for bulk JSC Mars-1 shown above.

Note that smaller-grained materials, such as the carbon particles employed in the KOSI Experiments[20], were not attempted due to concerns that they could be lofted into the tubing from the sample chamber and into the QMS inlet, potentially damaging the mass spectrometer.

Sources of Error

There are several potential sources of error for any particular D/H value to be reported. The largest uncertainty is in the partial pressures of the various species measured by the Quadrupole Mass Spectrometer

(QMS). Measurements are taken of the entire mass spectrum every minute which means that a large number of points are associated with any run. Since the rate of the change of the pressure is typically slow, it is possible to perform a statistical analysis on the individual measurements over a few hours to derive 1-σ errors for the D/H ratio, these are shown in Table 2and range from 0.7% to 6.9% of the total value.

Table 2: List of palagonite experimental runs

Run	Grain Size	Dust Content	Duration	Basal T	1- D/H Error	max fractionation
7	1-10 μm	1 wt%	7.8 days	60 K	5.2×10^{-4}	1.0
8A	1-10 μm	3 wt%	12.8 days	60 K	6.6×10^{-4}	1.2
8B*	1-10 μm	3 wt%	7.0 days	60 K	6.5×10^{-4}	1.1
9	1-10 μm	6 wt%	17.8 days	60 K	1.4×10^{-3}	1.2
10	1-10 μm	9 wt%	19.6 days	60 K	1.9×10^{-3}	1.9
11	1-10 μm	25 wt%	36.0 days	60 K	1.3×10^{-3}	2.7(1.6)
12	105-150 μm	9 wt%	15.6 days	60 K	3.7×10^{-3}	1.0

*Following Excavation of topmost layer

Moores *et al. Planetary Science* 2012 1:2, doi:10.1186/2191-2521-1-2

There could also be uncertainty in the preparation of samples; however, this is expected to be smaller. All liquid samples were prepared using calibrated volumetric pipettes. Furthermore, rapidly frozen mixtures of HDO and H_2O have not been found to fractionate significantly, especially when the freezing is rapid [34,35]. Nevertheless, the final values of the D/H ratio were compared to the initial value obtained at the start of the experiment during which the surfaces of the samples were unprocessed and should have been sublimating at close to their bulk isotopic ratios.

RESULTS

The experimental runs conducted are summarized in Table 2 which lists the grain size of dust, the dust mixing ratio, duration, basal temperature, the greatest 1- error in the D/H ratio, typically at the end of the run, and the maximum fractionation factor. The fractionation factor is the ratio between the bulk D/H ratio of the original solid and the observed minimum gas D/H ratio.

Typical Pressure and Temperature Profiles

A typical pressure and temperature profile from the dusty samples is seen in Figure 3, plotting the outputs of the 20 mtorr manometer and the Temperature sensor located at a height of 5 cm above the base of the sample chamber. This temperature sensor is located close to the sample surface, but likely is in poor thermal contact with the sample due to the high porosity of the sample. This likely explains the low temperature reported which should be between 181 K and 207 K, assuming saturation in the chamber and the pressure range seen in the upper panel.

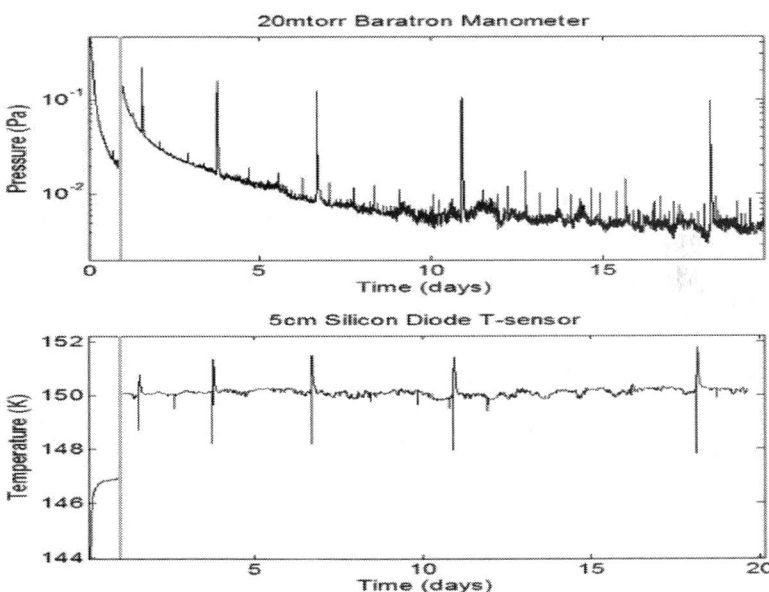

Figure 3: Evolution of the pressure at the 20mtorr manometer and the temperature sensor near the sample surface for the run described in Figure 4, panel E. The time of first illumination by the lamp is given by a vertical grey bar. Note the large decrease in pressure observed over the course of the run compared to a lack of an appreciable temperature change.

An important feature is the decline in pressure which continues for the duration of the experiment. This decline is accompanied by a stable temperature regime which suggests changes occurring in the sample itself. This type of a pressure decline has been seen previously in crystalline experiments [19] and is a well documented feature of the dusty KOSI experiments [20]. In the KOSI experiments, the pressure decline observed was assumed to be the result of the formation of a dusty lag layer [36]. This possibility will be discussed further in Section "Discussion".

For several reasons, it seems unlikely that the decrease in pressure is caused by the gradual removal of adsorbed water on the interior surfaces of the chamber and lines. First, the analysis line was maintained at a high vacuum at all times and was heated between runs to drive off any adsorbed water. Background measurements detected contamination from water at the level of only 2×10^{-8} torr, more than three orders of magnitude lower than the lowest pressure observed during any experimental run. Furthermore, since the sample chamber contains a water ice sample which is directly heated from above, it seems unlikely that the walls which contain only adsorbed water and were not heated, would contribute a greater signal by several orders of magnitude. But most tellingly, the D/H profile for such a process is opposite to what would be expected. The end of each sample run produced the most HDO-poor gas and it would be this gas which would be expected to remain on the surface of any plumbing. Next, during sample replacement, very HDO-poor room air was permitted to enter the sample chamber while the chamber was at room temperature. If either of these sources was contributing heavily to the D/H ratio of gas emitted from the chamber, one would expect the initial D/H ratio to be low and to increase as this interfering contamination became exhausted. Instead, the opposite trend is observed (see Sections "Palagonite series" and "Comparison to other dust compositions and blank runs").

Next, a discontinuity is observed at 1 day of elapsed time. This corresponds to turning on the Quartz-Tungsten-Halogen (QTH) lamp to provide heat directly to the surface of the sample. As expected, this event corresponds to sudden increases in the temperature and pressure. Immediately following this event, both signals return to their initial tendencies: a decline in pressure at the manometer, and a stable temperature signal. Since the temperature stabilizes within ten minutes, thermal equilibrium is achieved quickly.

The last significant feature is the presence of five large spikes in pressure and temperature at declining intervals. These large excursions are similar in structure to the events termed "sublimation cascades" in [19] and seen in several different experimental apparatuses. As in [19] we observe that these events are accompanied by excursions in the D/H ratio of the sublimating samples, however, their contribution to the overall proportions of HDO and H_2O sublimated is small and therefore can be neglected when considering the trends in the D/H ratio.

Palagonite Series

Overview

Figure 4 shows seven experimental runs in which the dust content varies from 1 wt% in panel A through 3 wt% in panel B and C, 6 wt% in panel D, 9 wt% in panel E and 25 wt% in panel F. Panel C demonstrates the effect of removing 1 cm of overburden from the surface of a sample which had been sublimating for 13 days. This removal was accomplished by opening the sample chamber, backfilling with LN2 and removing the upper part of the sample with a spatula and was completed within a few minutes. Panel G examines the effect of changing the size of dust particles, from 1-10 μm to 76-105 μm, while holding the mass concentration constant at 9 wt%. As dust concentration is described as a mass mixing ratio, this final experiment provided a way to separate the effect of particle size, and hence the specific surface area since larger particles have less surface area per unit mass, of the particulates on the sublimating ice.

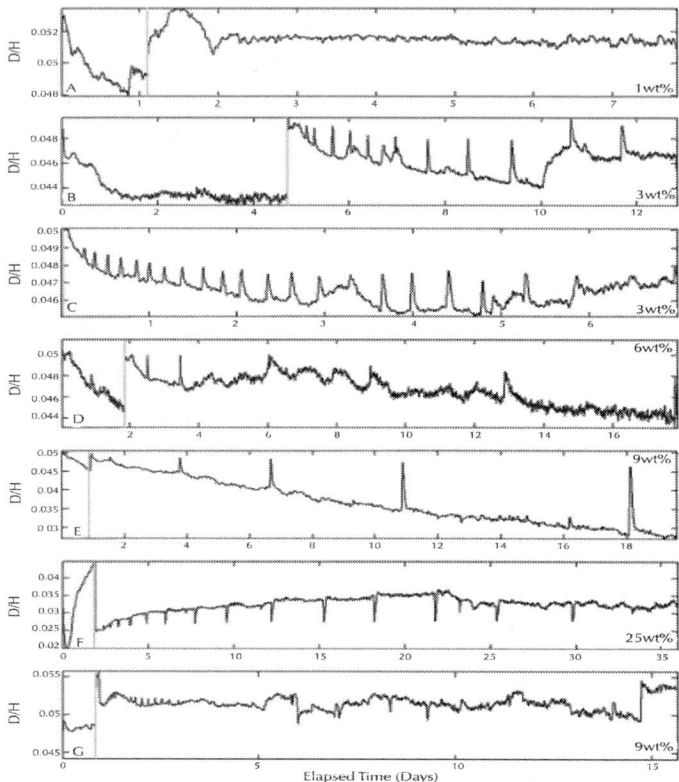

Figure 4: Palagonite series, note differences in the x and y scale between panels. The time of first illumination by the lamp is given by a vertical grey bar. From top to bottom 1 wt% 1-10 µm particles, 3 wt% 1-10 µm particles, 3 wt% 1-10 µm particles following excavation of the top 1 cm of sample, 6 wt% 1-10 µm particles, 9 wt% 1-10 µm particles, 25 wt% 1-10 µm particles, 9 wt% 76-105 µm particles Only dust contents 9 wt% and 25 wt% are heavily fractionated by the end of their runs, while all ices contaminated with fine dusts show some degree of increasing fractionation with time.

This last panel also illustrates the response of samples to changing illumination conditions. At day 1 and again just prior to day 15, the intensity of the lamp was increased as a step. The immediate response of the D/H ratio was to increase as well. This is also observed for all the other palagonites, particularly the lower wt%, but not for the 25 wt% sample which saw a decrease in the D/H ratio of the sublimate gas upon being illuminated.

In many cases, as mentioned in section «Typical pressure and temperature profiles», the initial pressure in the chamber exceeded the maximum pressure at the RGA. As such, it was necessary to wait until the pressure had declined sufficiently before beginning the illumination. This process took from a few hours to 5 days depending upon the particular sample in question. As a result there is an initial large jump in pressure which often shows up as a large jump in the D/H ratio within the first five days of each run that corresponds to this event, such as just before day 5 in panel B, day 2 in panel D, day 1 in panel E, day 2 in panel F and day 1 in panel G. A clear jump due to this effect cannot be discerned in panel A, and panel C represents a continuation of panel B--therefore it is fair to say that the jump occurs between the two runs. Note that the lamp is first turned on at 27 hours (1.13 days).

A Trend in Dust

From this series of experiments it is possible to discern a trend in the amount and speed with which fractionation of the sublimated gas takes place. The lowest dust concentration, 1 wt%, is shown in Figure 4, panel A, and shows almost no fractionation. Instead, the D/H ratio remains stable at 0.052, near the bulk ratio after an initial excursion following first illumination just after 1 day. At 3 wt% (panels B and C) it is possible to discern a slight downward trend in the D/H ratio bottoming out around 0.045 by the end of the run. Interestingly, there seem to be several fractionation reverses despite the appearance of somewhat stable pressure behavior. Most notably these occur at day 10 in panel B and around day 3 in panel C. As mentioned above, panel C describes the sample following the removal of the top centimeter of material. From the initial high D/H value at the start of the run shown in panel C it appears that, at depth, the sample remains unfractionated. However, once this surface is exposed it begins to fractionate just as the original surface had in panel B. At 6 wt% (panel D) the fractionation becomes slightly more pronounced, with D/H declining to 0.044 by the end of the run. At 9 wt%, the D/H value declines to 0.028 by the end of the run, a fractionation factor of 1.9. This run exhibits the largest end-time fractionation of any run completed.

After 9 wt% the dust content was increased to 25 wt%. This sample (shown in panel F) exhibited a different behavior compared to samples with lower dust contents. The sample began to fractionate heavily prior to the day 2 lamp illumination. This fractionation was reversed after peaking at a factor of 2.5, with the D/H ratio ascending up to 0.045. The lamp illumination, instead of being accompanied by a spike in the D/H ratio is instead shown as a sudden decrease of D/H down to 0.025. By the end of the run, D/H had increased only to 0.032, an apparent asymptote, sublimating during the entire run at D/H values much lower than the bulk. This suggests that this heavy fractionation may be a long term effect and not simply transitory. The specific elements of the 25 wt% case which differ from the 9 wt% case and a possible explanation for both are discussed more fully in the Section "Using adsorption to explain a special case".

The final panel of Figure 4 describes the effect of replacing the 1-10 μm particles of the 9 wt% case with 76-105 μm particles. The degree of fractionation is much less than the 1-10 μm case. The result is a run that shares many aspects of behavior with the 1 wt% case. This is suggestive that the fractionation may be related to the total surface area of dust available and not to the mass of dust.

Comparison to Other Dust Compositions and Blank Runs

Since experiments with JSC MARS-1 palagonite dusts were more extensive than those with other dusts, these have been used to illustrate the trend of increasing D/H fractionation with increasing dust content. However, it is noteworthy that similar D/H profiles were obtained from all dusts mentioned in the Section "Dust and regolith simulants". Furthermore, blank runs which contained no dust and which were prepared in an analogous manner to the runs which contained dust, showed no significant fractionation during runs of over 30 days. Both sets of observations are presented in Figure 5 with significant fractionation observed on SiO_2 with D/H values observed down to 0.03 (Panel C), and on TiO_2 with D/H values down to 0.02 (Panel D). Less fractionation was observed on the Pallasite cuttings (Panel E). This effect is possibly due to the larger size of these cuttings when compared to the palagonites, SiO_2 and TiO_2.

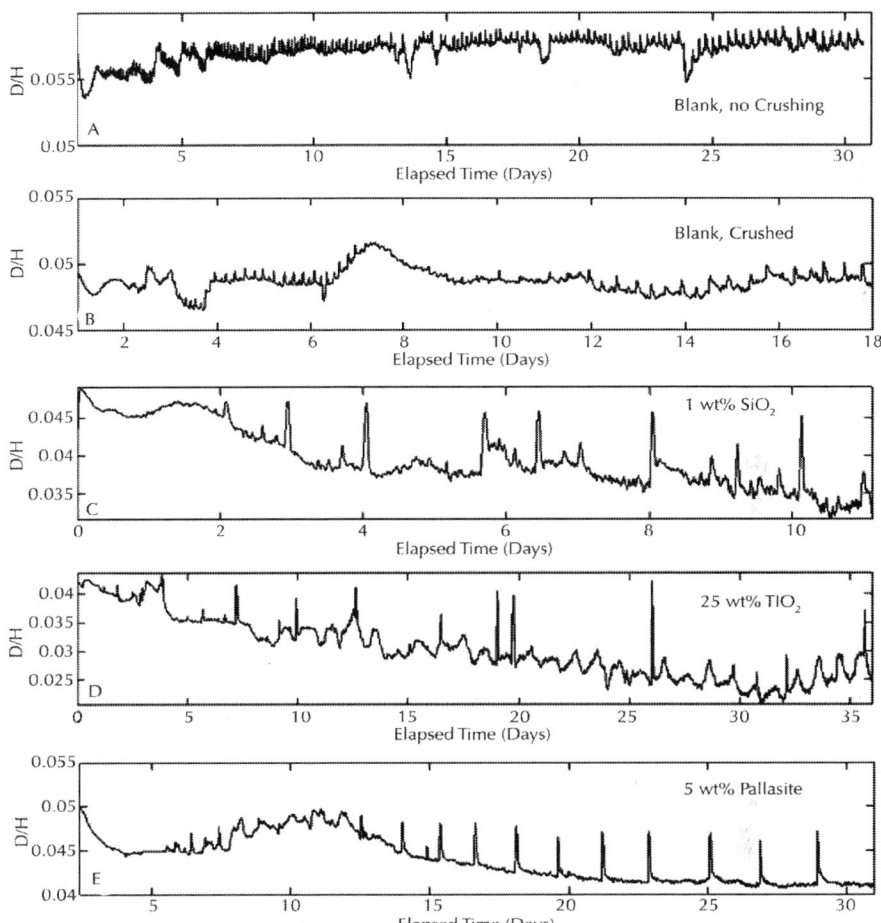

Figure 5: Blank Runs and other dusts. Lamp illumination occurred near the start time shown for all runs. (A) A blank (no dust) prepared by pipetting water into LN2 in the sample chamber or (B) pipetting water into LN2 outside the chamber and crushing with a mortar and pestle. Neither blank shows a significant fractionation; however, they give an idea of the degree of non-uniformity of disaggregated samples. Mixtures prepared analogously to panel B, but with different dusts are shown in the bottom three panels. These dusts are (C) 1 wt% SiO_2 in 1-2 μm particles (D) 25 wt% TiO_2 in 1-2 μm particles and (E) 5 wt% cuttings from the Fukang Pallasite in 1-50 μm particles. All three dusts show some increase in fractionation over time.

DISCUSSION

Pressure Decline

A Possible Cause: Dust Lag Formation

The decline in pressure observed in all samples over time is puzzling when considered in light of temperature readings which are essentially constant. Typically, for a sublimating sample, these two variables are highly correlated--in equilibrium sublimation, it is the temperature which uniquely determines the pressure in the sample head space. As such, starting pressures, ending pressures and the corresponding temperature of the sublimating surface, assuming a saturated equilibrium in the chamber, are given in Table 3. Graphically, a typical pressure decline is shown in Figure 3 along with the corresponding temperature trace for the 9 wt% palagonite.

Table 3: Pressure/temperature decay parameters for palagonite runs

Run	P_{START} (mtorr)	T_{SURF}^{*} (K)	P_{END} (mtorr)	T_{SURF}^{*} (K)
0**	11	216	0.53	195
7	2.5	205	0.21	189
8A	2.7	206	0.060	182
8B	3.0	207	0.18	189
9	3.5	207	0.049	181
10	2.9	206	0.040	180
11	4.8	210	0.034	179
12	6.6	212	0.043	181

*Based on Starting/Ending Pressure (Saturation Assumed) may not be representative of the actual temperature

**Slowly frozen ice from [20]

Moores et al. Planetary Science 2012 1:2, doi:10.1186/2191-2521-1-2

It is noteworthy that this kind of profile can also be seen in previous work. The KOSI-9 experiment in which the dust to ice ratio was 0.11 (9.9 wt%) is a particularly good example. Here, over a three hour stretch where the illumination was held constant, the flux of escaping particles increases, peaks and begins to fall before the lamp power is reduced [27]. Unfortunately, the lamp output does not remain constant for long enough to determine the nature of this pressure profile.

Thus, it seems likely that the increased surface area of the disaggregated samples and possible reorganization of pathways through the sample may explain the larger scale of the declines. The most straightforward way for a dusty sample to produce these pathways is through the formation of a dust lag. Such a lag creates a diffusive barrier to sublimation which can lead to a decline in the observed sublimation rate. This is what had been assumed to have occurred in the KOSI simulations [27] and was modeled by [36]. As such, the pressure decline that would be expected is documented in the following section and our results are compared to those of [36].

Quantifying Putative Dust Lag Formation

Before determining how great a barrier this dust mantle is to escaping particles, it is necessary to calculate the amount of dust lag that is expected from a sublimating sample. Typical rates of sublimation seen in dusty runs are of order 10 μm day^{-1} of clean ice equivalent. If it is assumed that the excavated dust is loosely packed with porosities approaching 50%, typical for bulk JSC Mars-1 [30] then the thickness of lag creation should vary from 0.08 μm day^{-1} to 2.7 μm day^{-1} for 1 wt% to 25 wt% dust mixing ratios, respectively. The rate of lag creation will vary over time, thus it is useful to look at the total lag built up by the end of the run. This is estimated based on the total mass of material sublimated and tabulated in columns 2 and 3 of Table 4.

Table 4: Dust lag and diffusion timescales for 1-10 μm palagonite series

Run	H$_2$O$_{Eq}$ Layer	Estimated Lag	C$_{KOSI}$	Mantle	R^2	Effective Pore
		Thickness	**(m kg^{-1} s$^{-0.5}$)**	*Diffusivity*		*Radius**
7	536.8 μm	6.78 μm	1.3 × 10^{19}	2.0 × 10^{-7} m^{2}s^{-1}	0.95	2.8 × 10^{-10}m
8A	421.3 μm	16.3 μm	2.8 × 10^{18}	2.9 × 10^{-8} m^{2}s^{-1}	0.76	4.0 × 10^{-11}m

8B	325.4 μm	12.5 μm	1.6×10^{18}	$8.9 \times 10^{-9} m^2 s^{-1}$	0.79	$1.2 \times 10^{-11} m$
9	408.8 μm	32.6 μm	1.9×10^{18}	$2.5 \times 10^{-8} m^2 s^{-1}$	0.77	$3.5 \times 10^{-11} m$
10	250.2 μm	30.9 μm	1.2×10^{18}	$1.5 \times 10^{-8} m^2 s^{-1}$	0.67	$2.1 \times 10^{-11} m$
11	547.0 μm	228 μm	2.7×10^{18}	$2.1 \times 10^{-7} m^2 s^{-1}$	0.42	$2.9 \times 10^{-10} m$
12	1017 μm	126 μm	5.5×10^{18}	$3.3 \times 10^{-7} m^2 s^{-1}$	0.75	$4.5 \times 10^{-10} m$

*Assuming that the pressure decline can only be explained without considering adsorption

Moores et al. Planetary Science 2012 1:2, doi:10.1186/2191-2521-1-2

The speed of diffusion of water molecules will also be affected by the temperature of this dust lag. For this variable, it will be assumed that the lag does not vary in temperature over its thickness and is the same temperature as the sublimating surface. This is reasonable given the estimates of lag thicknesses which are less than 228 μm in all cases at all times. This thickness when combined with reasonable assumptions regarding the thermal conductivity suggest that the temperature drop across the lags will be less than 0.01 K.

Using this information, the thickness of the lag at any time (and the corresponding chamber pressure) can be calculated by solving the following equations. First, Fick›s first law for the molecular flux (molecules $m^{-2} s^{-1}$) in terms of pressure can be expressed as:

$$J|_{diffusion} = -D_{eff} \frac{\partial}{\partial z} \left(\frac{p}{m_{H2O} RT} \right) \cong \frac{-D_{eff}}{m_{H2O} TR} \frac{\Delta p}{\Delta z} = A \frac{(p_{sat} - p)}{\Delta z}$$

(1.a)

Where the temperature has been separated from the derivative as the temperature drop across the lag layer is expected to be small and m_{H2O} refers to the mass of a single water molecule. D_{eff} is the observed diffusivity in $m^2 s^{-1}$, z is the lag thickness in m, p is the pressure in Pa, T the temperature in K and R is the specific gas constant for water of 462 $m^2 K^{-1} s^{-2}$. All constant terms are absorbed into A for simplicity, and p_{sat} is the vapor pressure of the ice at the actively sublimating surface below the lag. Note that all variables used in this analysis and all the analysis to follow in this paper are defined in Table 1. As water molecules are evaporated, the lag will thicken according to:

$$\frac{dz}{dt} = J \frac{M_{H2O}}{N_A} \frac{\chi z}{\chi_{H2O}} \frac{1}{\rho z \gamma z} = BJ$$

(1.b)

Where M_m is the molar mass of water, N_A is Avogadro's number, χ_Z and χ_{H2O} are the mass fractions of dust and water, respectively, ρ_Z is the density of dust, and γ_Z is the proportion of the volume of lag that is dust, as opposed to open space. As before, constant terms are absorbed into the constant B. Finally, using the parameters of the setup, we may relate the flux of escaping particles to the pressure in the chamber by:

$$p = J \frac{k_B T_{QMS} A_{surf}}{e_{sys} \dot{V}} = CJ$$

(1.c)

Where k_B is Boltzmann's constant, T_{QMS} is the temperature of the tubing at the throat of the QMS, taken to be equal to the room temperature of 295 K, A_{surf} is the surface area of the sample, e_{sys} is the efficiency of the system (the ratio between the pressure at the throat of the QMS and in the chamber, nearly a constant 0.01 over the pressures involved) and the final constant is the pumping speed of the system, a nearly constant $0.02 \ m^3 \ s^{-1}$ (20 L s^{-1}). Note that in the limit of a system with completely free sublimation the back pressure tends towards zero.

Solving these three equations yields expressions for the escaping flux of particles, chamber pressure and the lag thickness in terms of t, the time elapsed:

$$z(t) = \sqrt{A^2 C^2 + z_0^2 + 2ACz_0 + 2p_{sat}AB(t - t_0)} - AC$$

(2.a)

$$p(t) = \frac{ACp_{sat}}{\sqrt{A^2 C^2 + z_0^2 + 2ACz_0 + 2p_{sat}AB(t - t_0)}}$$

(2.b)

$$J(t) = \frac{Ap_{sat}}{\sqrt{A^2 C^2 + z_0^2 + 2ACz_0 + 2p_{sat}AB(t - t_0)}}$$

(2.c)

Where z_0 is the thickness of the lag at $t = t_0$ and accounts for any lag which has accrued from insertion of the sample and chamber pump-down. As t becomes large compared to the initial conditions, the chamber-specific constant, C, disappears and the flux of particles approaches:

$$J(t) = \frac{Ap_{sat}}{\sqrt{2p_{sat}AB(t - t_0)}}$$

(3.a)

This form is similar to the decrease in pressure due to dust mantling observed during the KOSI experiments and modeled by [36] whose expression for the escape rate can be given as:

$$J_{KOSI}(t) = \frac{C_{KOSI}\rho}{2\sqrt{(t - t_0)}}$$

(3.b)

For these experiments, [36] found that on a mass-basis C_{KOSI} had a value of 2.3 × 10⁻⁵ at an input flux of 1300 Wm⁻² and 0.5 × 10⁻⁵ at 320 Wm⁻² where the particle flux, J_{KOSI}, is given in units of 10⁻⁴g cm⁻² s. In terms of molecules and mks units, these constants are 7.7 × 10¹⁸ molecules m kg⁻¹ s⁻⁰·⁵ and 1.7 × 10¹⁸ molecules m kg⁻¹ s⁻⁰·⁵. Most of the values obtained by curve fitting fall into this range with values of C_{KOSI} ranging from 1.2 to 13 × 10¹⁸ molecules m kg⁻¹ s⁻⁰·⁵. The results of these fits, along with the goodness of fit R^2 parameter (Coefficient of Determination) are documented in Table4. Additionally, using the derived KOSI constant and equations (1.a), (2.c) and (3.a) an equation for the diffusivity of the dust mantle may be derived:

$$D_{eff} = m_{H2O}TR\frac{B\rho^2 C_{KOSI}^2}{2p_{sat}}$$

(4)

The values for the diffusivity of the dust lag derived in this way are also documented in Table 4 and range from 9 × 10⁻⁹m²s⁻¹ to 3 × 10⁻⁷m²s⁻¹ and average 1.2 × 10⁻⁷m²s⁻¹. These values are too small to represent molecular diffusion through the chamber, which is of order 10⁻³m²s⁻¹ at these temperatures and pressures [37].

These diffusivities can be used to determine the average spacing of the particles by employing calculations for Knudsen diffusion through cylindrical pores between particles in the dust mantle using the formulation of [38]. These calculations give a characteristic pore size of less than an angstrom, smaller than a water molecule. However, since the particles which make up the dust lag are in actuality only a few microns in size, it was expected that pore sizes would be of similar size. Performing the Knudsen diffusion calculations for micron-sized pores gives values that are similar to the values for molecular diffusivity, namely 10⁻³m²s⁻¹. While significant tortuosity could reduce this figure, it is doubtful that sufficient tortuosity exists within a dust mantle only tens of particles thick to reduce the observed diffusivity by four orders of magnitude.

Thus, it seems unlikely that dust mantling alone is responsible for the reduction in pressure that is observed within the sample chamber. Another possibility for this trend is given by adsorption which has previously been seen to produce abnormally low effective diffusivities under certain conditions[39-41]. More evidence for this possibility comes from a consideration of the large fractionations observed. Therefore adsorption will be considered in greater detail in the Section "Dust mantle adsorption".

Heavy Fractionation during Sublimation

Discussion of Cold Trapping

Why is it that the D/H ratio of the sublimate gas decreases with time? This effect is not seen in solid samples. Depending on the rate of solid state diffusion within the sample and the rate of removal of material from the head space, solid samples may sublimate un-fractionated, i.e. at the bulk ratio of the solid as monolayer after monolayer are stripped away. Or, instead, even solid samples may exhibit a Rayleigh-like fractionation in which the head space gas becomes more isotopically heavy with time e.g. [19] if diffusion allows molecules to travel away from the actively sublimating surface faster than the sublimation front proceeds.

However, solid samples do not have direct pathways down to a cold trap which can have a temperature as much as 150 K less than the actively sublimating surface. Thus, cold trapping of HDO is an attractive possibility for producing an overabundance of the light isotope in the headspace gas. In light of the large temperature gradients that exist in the samples under consideration and the corresponding high fractionating factors predicted by the kinetic isotope effect, this mechanism for enriching the headspace gas in the light isotope must be examined.

Some insight can be gained by considering the results of the KOSI experiments along with simulations performed by [42]. In both cases, it was observed that condensable material evaporated from the surface would penetrate into the subsurface. However, due to the decrease in temperature this material recondensed quickly before travelling deeply into the sample. As such, a short distance below the surface, a peak was seen in the diffusing material, followed by a drop-off to normal conditions in the deeper interior. This suggests that volatile transport will be localized to a layer near the surface, and that the concentration of the more refractory component, HDO, will increase at the surface due to preferential transport of the lighter, more volatile H_2O. This is exactly what has been observed in the KOSI experiments where the D/H ratio in the near surface was seen to be slightly enriched in deuterium by 47.7‰ at the end of the 34-hour KOSI-7 run [29] while the rest of the sample was unaffected.

A simple numerical simulation shows that the samples considered in this paper are no different. By altering Eq. (1.a) such that the temperature is allowed to vary and applying it to the sample interior instead of the dust mantle, the rate of diffusion and flux of H_2O and HDO through the sample can be calculated based on some simple assumptions about the geometry and kinetics. For the samples considered in this paper, a pore radius of at most 0.5 mm, corresponding to the typical size of the crushed ice-dust grains, and a temperature gradient of 30 K cm^{-1} are reasonable, which implies utilizing the Knudsen diffusion equation as the pore radius is smaller than the mean free path at these temperatures. The results are shown in Figure 6. As can be seen, the rate of transfer becomes relatively small below one centimeter from the surface. This perhaps explains why the D/H ratio appears to have been reset in the dusty runs where the top 1 cm was removed between runs (Table2 runs 8A and 8B).

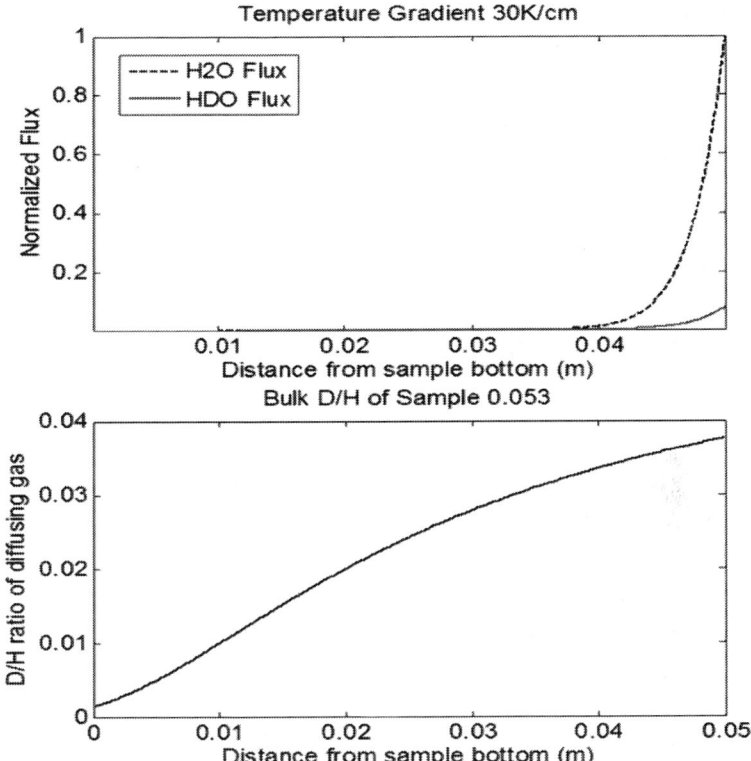

Figure 6: Numerical model results showing the normalized rate of gas diffusion at different levels and the D to H ratio of the diffusing gas in the sample at the beginning of a run with a surface temperature of 210 K and a thermal gradient of 30 K/cm. Total depth is 5 cm (0.05 m). Below 1 cm from the surface, the fluxes become very low, restricting the flow of material. As such, (1) large amounts of gas will condense out before reaching the highly fractionating bottom layers. Also, (2) since condensation greatly exceeds sublimation at depth, the inward-travelling gas is isotopically light, composed principally of H_2O.

The most significant argument against implicating pathways in the fractionation process is that a trend in dustiness would not be expected for this phenomenon. There are runs when the pathways to the cold finger are present with very little dust or in the absence of dust entirely. The top panel in Figure 4 is a good example. In none of these cases was a significant fractionation observed.

Dust Mantle Adsorption

As such, the dust must be the source of the fractionation observed. This is supported by the observation that fractionation proceeds more rapidly when more dust is entrained in the sample. The major way in which this dust can interact with the gas and ice in the chamber is through adsorption. Previous work [39-41,43,44] on the palagonites used to simulate martian surface materials, such as JSC MARS-1 indicate that palagonites should readily interact with water vapor at low temperatures. This would explain the very low effective diffusivities observed in the Section "Pressure decline" without having to resort to high tortuosities or extremely tiny pore radii.

This behavior has been previously noted in ice sublimating through regolith [39,40] and can result in an observed effective diffusivity that is much lower than would be predicted by either molecular or Knudsen diffusion alone. The situation discussed in [40] is not dissimilar from a sublimating dusty ice which is forming a dust mantle, as is occurring in the experiments detailed in this paper. Thus, it is useful to consider the analysis from [40]. From Chevrier's equation (15) the reduction in the diffusivity can be expressed as:

$$D_{eff} = \frac{D_0}{1 + \psi \dfrac{\partial \theta}{\partial p}}$$

(5)

Where D_{eff} is the effective or observed diffusivity of the dust or regolith layer, D_0 is the diffusivity controlling the movement of vapor between dust particles, i.e. the lesser of molecular and Knudsen diffusivities for the given situation. ψ is a thermodynamic constant defined by [40] as equation (22) repeated below:

$$\psi = \frac{R_u T \rho_{H2O} \rho_{reg} A_s l}{M_{H2O}}$$

(6)

Where R_u is the universal gas constant, T is again the surface temperature of the sublimating ice, M_{H2O} is the molar mass of water, 0.01802 kg mol^{-1}, l is the thickness of an adsorbed monolayer, 2.75×10^{-10} m, ρ_{H2O} and ρ_{reg} are the densities of the adsorbed water layer and the regolith, respectively, 1000 kg m^{-3} and 800 kg m^{-3} and A_s is the specific surface area in m²kg⁻¹. For the situation described in this paper, ~ 1.4×10^7 J m⁻³.

The last element from equation (5) is the slope of the ϑ-p relationship where   is the surface coverage and p is the pressure of water vapor in contact with the regolith. The formulation of this term is slightly different from [40] equation (23) in that here particles are exhumed from solid ice by sublimation instead of starting out as dry overburden. They can therefore be considered to start out with completely filled monolayers of adsorbate under a Langmuir scheme, i.e $\vartheta(t = 0) = 1$. This implies (see [40] for a description of the derivation):

$$\frac{\partial \theta}{\partial p} = \frac{\alpha}{(1 + \alpha p)^2}[1 - (k_D t(1 + \alpha p) + 1)e^{-k_D(1+\alpha p)}]$$

(7.a)

Where k_D is the desorption rate constant and α is the adsorption coefficient. This equation contains the time-varying terms which have an effect on the diffusivity when the state of adsorption in the dust mantle is changing. These are particularly important for the case of dry overburden where the Langmuir isotherm is steep and the slope of the ϑ-p relationship is large and rapidly evolving as more and more water vapor becomes adsorbed on the surface. However, for the case in which material begins fully saturated, large reductions in the surface coverage, ϑ, can be achieved without changing the slope of the ϑ-p curve significantly. Additionally, since the limiting behavior of the system is of greatest interest, it is useful to consider the system at $t >> t_0$. Both of these criteria suggest using the following simplified form for the adsorption slope in which equation (7.a) reduces to:

$$\frac{\partial \theta}{\partial p}\bigg|_{t>>0} = \frac{\alpha}{(1 + \alpha p)^2}$$

(7.b)

This formulation is nearly entirely time-independent, having only a dependence on pressure. Since the effective diffusivities average close to 10^{-7} m² s⁻¹, and the molecular and Knudsen diffusivities for particles of this size are of order 10^{-3} m² s⁻¹, the value of the denominator in equation (5) is of order 10^4. This implies values for the slope of order 10^{-3} m³ J⁻¹ and values of α of either $> 10^3$ or $< 10^{-3}$ for the pressures encountered in the experimental apparatus, and the value of ψ.

Only the larger set of these values is physical. To explain this assertion, consider that the adsorption coefficient may be expressed as the ratio of adsorptive and desorptive rate constants. At low temperatures, adsorption should be more effective and desorption should be less effective. Thus, it is expected that values for α will be several orders of magnitude greater than at higher temperatures, such as at 270 K where $\alpha \sim 0.01\text{-}0.1$ is a more common range [40] or at 243 K where $\alpha \sim 0.81$ [41]. Since α cannot be less than 0.81 at 200 K, the higher values are more appropriate.

Adsorptive Fractionation

When HDO is introduced, the situation becomes more complicated. In addition to the adsorption and desorption processes for each species, it may be necessary to consider the interactions between the two species [39] considered the effect of a two-component system on the adsorption and desorption process in their section 4.5 and show the results expected for a system with an isotopic composition of 1 HDO: 1000 H_2O in their figure nine. They observe that significant increases in the relative surface coverage of HDO are possible whenever α_{HDO} is greater than α_{H2O}.

How will this increase in the amount of HDO adsorbed on the dust mantle affect the composition of the sublimating gas? To determine what an idealized system with an adsorptive dust lag would look like, a numerical model was created using equations (1.a) through (1.c) to simultaneously solve for the pressure of HDO and H_2O in the headspace as the dust lag thickens. These dust lag differential equations were linked to adsorption through the effective diffusivity by making use of equation (5) modified for the two component system as:

$$D_{eff} = D_0 \frac{1}{1 + \psi \left(\dfrac{\partial \theta}{\partial p_{H2O}} + \dfrac{\partial \theta}{\partial p_{HDO}} \right)}$$

(8)

This equation, in turn, requires information on the adsorption state, , of the dust lag. If it is assumed that for each thickness of the dust lag a quasi-steady (i.e., time independent adsorption at each time step) state exists in terms of the adsorption state of the dust lag, then, for the 2-component system described here, the adsorptive state at each level within the dust may be described (after [39]) by:

$$\theta = -G^{-1}k$$

(9)

Where:

$$G = \begin{bmatrix} -k_{H2O}^{ads} - k_{H2O}^{des} - k_{HDO}^{X} & k_{H2O}^{X} - k_{H2O}^{ads} \\ k_{HDO}^{X} - k_{HDO}^{ads} & -k_{HDO}^{ads} - k_{HDO}^{des} - k_{H2O}^{X} \end{bmatrix}, \; k = \begin{bmatrix} k_{H2O}^{ads} \\ k_{HDO}^{ads} \end{bmatrix}$$

(10)

Note that the adsorption and desorption rate constants, k, are subscripted with the species in question and superscripted as to whether they represent rates for adsorption of a molecule on an empty site (ads), desorption of a molecule to leave an empty site (des) or the displacement of a molecule of one species by that of another (X).

If it is assumed that the temperature of the sublimating ice surface and the entire dust lag remains constant, then the desorption rate constants, k^{des}, for each component should themselves be constant. However, the adsorption coefficient will vary with pressure throughout the lag [after [40], equation (18)] according to:

$$k_{HXO}^{ads} = k_{HXO}^{des} \alpha_{HXO} p_{HXO}$$

(11)

Where the subscript HXO denotes either of H_2O or HDO. Here the desorption coefficients, k^{des}, are assumed to be equal for both species and the values used are those from [43] of approximately $1 \times 10^{-4} \, s^{-1}$. Next, α_{H2O} is known from equation (7.b) and α_{HDO} can be calculated from the relationship derived by [39] of $\alpha_{HDO} = \alpha_{H2O}/0.6$. Finally, for simplicity, all the substitution adsorption coefficients, k^{X}, are set to zero. The information from (9), (10) and (11) can be substituted back into equation (8) in order to determine the slope of the adsorption isotherm (slope of the ϑ-p curve).

Using this information, a fourth order Runge-Kutta (RK4) scheme was implemented to solve for the pressure of HDO and H_2O within the dust lag as a function of time. The pressure at the ice surface was assumed to be saturated at all times for both H_2O and HDO and the value of all kinetic constants were as described above. Runs were completed at several different dust concentrations ranging from 1 wt% up to 25 wt%. A value of 7000 for α_{H2O} was selected as the time required for fractionation in the simulations using this value matched reasonably well with what was observed in Figure 5 and it satisfies the requirements of equation (7.b).

The result of this calculation, up to an elapsed time of 10 days is shown in Figure 7. The curves in Figure 7 replicate many of the features of the simulations described in the Section "Results". All show a decrease in the D/H ratio of the sublimating gas with time. Also, the higher the initial dust content, the more rapid is the rate of decrease. However, no matter the starting concentration, the end fractionation is the same, with the system asymptotically approaching a D/H value of:

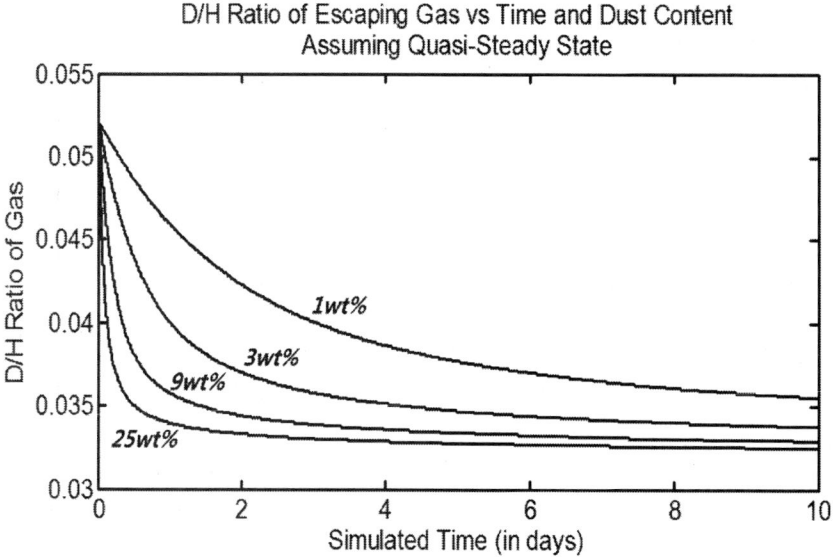

Figure 7: Numerical Model runs describing the D/H ratio of the sublimating gas as a function of time and dustiness for the H_2O-HDO system and starting concentrations relevant to the experiments documented in Figure 5. Substitution is assumed negligible. The value of α_{HDO} is set at $\alpha_{H2O}/0.6$. The dustier the original starting material the faster each curve approaches the limiting value of D/H = $\alpha_{H2O}/\alpha_{HDO}$ = 0.6 $(D/H)_0$. Different values for α_{H2O} will produce different timescales for reaching the asymptotic D/H ratio. A reasonable value, 7000, is shown here.

$$\lim_{t \to \infty} \frac{D}{H} = \frac{D}{H}\bigg|_{t=t_0} \frac{\alpha_{H2O}}{\alpha_{HDO}}$$

(12)

What does it mean for α_{HDO} to be greater than α_{H2O}? This suggests that at the same temperature, the affinity of the dust for HDO is higher than for H_2O. Since H and D have been observed in other experiments to fractionate heavily on surfaces [45,46] at low temperature, it is reasonable that HDO could be adsorptively sticky, compared to H_2O. In the case of H_2 and HD this assumption is a common one i.e. [47] and is often a starting point in considering exchange on the surfaces of dust grains that further sequester the deuterated species.

As far as fractionation is concerned, a larger α_{HDO} implies a smaller D_{HDO}^{eff}. This means that as gas passes through the dust, HDO is impeded more than is H_2O. As a result of this adsorptive separation, a stable fractionation can be obtained between the two gasses, as described by Eq. (12). Adsorption provides an effective mechanism for achieving this fractionation due to the gradient that will exist within the dust lag layer. Newly exhumed particles emerge with all adsorption sites filled, but over time this material is lost. Since the lag particles remain at the same temperature, this cold dust readily adsorbs molecules attempting to transit the layer. Later on these molecules are released through desorption in a mechanism similar to scattering. This adsorption-desorption process increases the transit time of molecules through the layer, giving rise to the very high apparent diffusivities observed. The closer the molecules are to the surface of the lag, the fewer adsorption sites are filled and the greater is this effect. Over time, as the lag thickens, the gradient in ϑ increases until, ultimately, the pressure of each species in the headspace reflects the affinity of the lag for each species rather than the difference in sublimation rates at the solid ice surface.

Finally, a note of caution is required. The apparatus and experimental method employed was chosen primarily for its ability to represent a realistically sublimating dusty surface and included all the imperfections inherent in such an experiment. It was not selected for the ability to precisely examine the kinetics of adsorption and to provide specific constants at low temperature.

Using Adsorption to Explain a Special Case

The framework developed in sections "Dust mantle adsorption" and "Adsorptive fractionation" can be used to explain the major differences in behavior between the 9 wt% run depicted in Figure 5E and the 25 wt% run depicted in Figure 5F. While both tend towards similar values for the D/H ratio in the long term, Figure 5E shows a declining trend after lamp illumination whereas Figure 5F shows an increasing trend. To understand how both cases can be explained by an adsorptive framework it is necessary to consider the cold illumination-free state before the initial lamp-on state. Initially, the ice begins by sublimating near the bulk D/H ratio. However, as dust is exhumed from the ice-dust matrix, water that has been adsorbed to that dust at the bulk ratio attempts to reach a new equilibrium. HDO is preferred by the dust and therefore replaces H_2O on dust grain surfaces and which is manifested as a progressive decline in the D/H ratio of the gas as the dust lag builds and scavenges HDO from the gas phase. This trend is seen in Figure 5F and several other runs including those depicted in Figures 5B, D and 5E prior to lamp illumination. In those three last runs, the process is short circuited at this point by the illumination of the lamp.

But for the run shown in Figure 5F, so much dust lag is exhumed prior to lamp illumination that the adsorption state of the dust evolves further. Eventually, the dust lag grows thick enough that it is does not only provide surfaces for adsorption, but is able to control the flow of gas from the subsurface. At this time a local equilibrium state is achieved at each level within the dust lag. As the gas in the headspace is HDO-poor compared to the ice surface, more and more HDO must be adsorbed on dust the lower that dust is within the lag, i.e. the closer that dust is to the ice. This acts to sequester HDO close to the ice layer.

However, this situation cannot persist. As the concentration of HDO increases at the base of the lag layer, the equilibrium at each level shifts towards greater and greater concentrations of HDO in the gas phase until, ultimately, the emitted gas from the top level has the same concentration as the gas supplied to the bottom of the dust lag. Thus, in the absence of illumination eventually a steady state is achieved with sublimation occurring at the bulk ratio. This typical transitory behavior of adsorptive systems is described well by [40] and was also seen in [39] and occurs when a stable dust lag or regolith evolves to a steady adsorption state throughout.

What happens when the lamp is illuminated? Immediately, the volume of gas sublimating from the ice surface increases by over an order of magnitude due to the sudden increase in temperature. But since the lag layer is already fully developed, of order 100 µm in thickness, the backpressure causes this ice to sublimate at the equilibrium ratio given by the Kinetic Isotope Effect (α ~1.3 at these temperatures [15]). After passing through the lag layer, the gas is further fractionated by adsorption (α = 1.7 at these temperatures [39]) which results in the first emitted gas having a D/H ratio of ~0.025.

It is the lamp that keeps the top of the lag layer in a low-ϑ state that allows it to continue to fractionate gas. Over time, the D/H ratio of the gas emitted from the top of the dust lag will increase until only the fractionation from the dust is present. The advancing sublimation front within the solid ice provides new material to the system at the bulk ratio. Thus the maximum fractionation is the dust lag fractionation multiplied by the bulk ratio. Interestingly, this implies that the D/H ratio at the dust lag/ice interface is larger than the bulk ratio by the amount specified by the kinetic isotope effect (i.e. 0.069 at the top of the solid ice layer).

The D/H ratio of the sublimation cascades suggests that this layer of enriched ice is thin compared to the thickness of ice that is consumed by a cascade. This can be shown by assuming that a significant amount of material is rapidly vaporized. If the enriched layer is small, on the average, this material will have close to a bulk composition. But since it is trapped underneath the dust lag by significant backpressure, HDO will recondense until the gas plug D/H ratio is equal to the value given by KIE. This gas then makes its way through the dust lag and is fractionated further, emerging with a D/H value close to 0.025 which is close to the peak values of D/H observed for these cascades.

Implications for Sublimating Water Ice on Icy Bodies

General implications

For bodies containing significant amounts of dust, preferential adsorption of HDO means that the solid is able to hold onto more deuterium than is predicted by the Kinetic Isotope Effect [48]. As a result, more fractionation can take place between the ice reservoir and the sublimate gas. Therefore the observed concentration of HDO in the sublimate gas will be lower than expected, or the source material will be richer in HDO than expected, depending on which planetary reservoir is being observed. In either case, the key to determining the significance of this experimental result on our understanding of the history of water in the solar system hinges on our knowledge of the plausible cycling between different reservoirs in a planetary setting and the pressure and temperature conditions under which the cycling occurs.

There are a great number of icy bodies in the solar system, but of these, this effect is likely to be significant mainly for two examples: Comets and the Martian polar caps. While the icy moons certainly contain large amounts of water ice, the vapor pressures involved are too small to mobilize this material, even over the timescale of the solar system, and their surfaces do not suggest large amounts of contaminating dust. Comets outgas heavily as they pass through the inner solar system and are known, from spectroscopy and space missions e.g. [49-51], to contain a great deal of dusty material. The polar caps of Mars are also known to be rich in dust [52] with up to 6% by weight on the surface, as determined by the OMEGA instrument on Mars Express [53] and potentially more at depth.

Comets

In some ways, the experiments discussed in this paper are a good analogue of a comet. Porosity estimates of the water ice grains produced in our experiments were, typically, near 40% or higher which is consistent with the data on cometary nuclei [54]. As well, the temperatures simulated in the lab were analogous to surface temperatures seen on comets, with back-pressures of about a Pa or less. The major differences between the experiment and reality are twofold. First the typical temperature gradient of 30 K cm^{-1} is likely far higher than what can be expected on comets. This means that, in contrast to our results, it is possible that there could be significant cold trapping if vapor is able to penetrate to depth within the comet. As well, comets likely have a layer of dry overburden much thicker than the fraction of a millimeter present by the end of the experimental runs described herein.

Comets are thought to represent early solar system materials, perhaps being made up of ice originally condensed from the solar nebula which has not since been processed or transported to another reservoir. As such, the isotopic composition of comets should be indicative of material present in the early solar system. It is in light of this information that the high value of the D/H ratio for most cometary coma, twice VSMOW [8-10,55], is puzzling. This value is enriched significantly compared to most other solar system reservoirs and represents an enrichment of almost 15 times compared to inferred protosolar values. Much of the work to date has, understandably, focused on attempting to explain how the observed enriched values of D/H in the coma could imply a much lower HDO concentration in the nucleus i.e. [56]. One of the main questions driving this debate is the source of water on the terrestrial planets. If comets are enriched in HDO compared to the earth, it becomes very difficult to form a set of circumstances under which a significant fraction of the Earth's water can be delivered by comets [1]. Recent observations by [55] of a D/H value near 1.0 times VSMOW in comet 103P/Hartley 2 has drawn this debate into sharper focus.

The results obtained from our sublimation experiments in the laboratory offer a potential solution. Since it seems likely that the heavier isotope is being concentrated on the dust, different values for the D/H ratio will be obtained depending on the provenance of the water vapor used for analysis. If that water vapor has been desorbed from dust grains, it is likely to have a higher D/H ratio than solid buried cometary ice. This will be especially true near the start of any degassing cycle before the comet has reached perihelion. Thus the D/H values obtained for comets in flybys of comae may not be at all representative of the reservoirs from which those comets formed depending on the stage of outgassing at which those comets are observed.

In order to better constrain the fractionation processes occurring on and near comets both spacecraft and ground-based observations are necessary [19] showed that the D/H ratio in sublimated gas can vary with distance from the cometary nucleus, increasing with distance. One possible reason for this trend might be that high-D/H material adsorbed on dust grains is evaporating as the grains are progressively dehydrated in their passage away from the comet. In order to understand this process, a spacecraft capable of measuring D/H would ideally provide measurements along a profile from the outer coma all the way to the surface of the nucleus.

Ground-based spectroscopy would also be helpful. To date, seven comets have been observed and their D/H values measured [55] but each was observed for only a few nights. While these measurements varied from 37 days before perihelion for comet Hyakutake [9] to 27 days past perihelion for comet 2002 T7 (LINEAR) [57] no comet has yet been observed over a long period which might show an evolution in the D/H ratio of the water vapor contained within the coma. D/H ratios within those comets vary significantly with six clustered near 1.9 times VSMOW and one near 1.0 times VSMOW [55]. This variation is comparable to the variation that can be imposed by adsorption on dust as discussed in the current paper. Therefore, long-term measurements of the evolution of the water in the coma could help to determine whether these snapshots of D/H are representative of the water in the nucleus.

Mars

Unlike a comet, the history of Mars contains numerous transfers between different reservoirs. As such, there is an extensive literature studying the D/H history of the planet in terms of plausible volatile cycling histories [58]. All routes from these ancient sources of water lead to the present-day polar caps where most of the modern surface ice currently resides [52]. Even here, the cap is marked by numerous layers and unconformities recording many advances, retreats, and additions to the ice cap which may be linked to past climate cycles [59]. These ice caps are themselves highly enriched in dust [52]. This dust is most likely similar in character to the atmospheric particles which cover the planet. These particles average 1.6 microns in radius as determined from the Imager for Mars Pathfinder [32] and Mars Exploration Rover missions [31] and are similar in spectra and composition to JSC Mars-1 analog, the dust simulant used in our experiments.

An additional advantage when considering Mars is the availability of fossil D/H ratios from martian meteorites. The oldest of these, the nahklites, chassigny and ALH 84001, all show relatively low D/H ratios of between 1 and 2 times VSMOW [60]. Since these meteorites are basaltic, it is presumed that the trapped water represents the primordial water of the Martian mantle which is likely representative of the initial D/H ratio of Martian water in general. Furthermore, the overall composition of the trapped water in QUE94201, a more recently formed shergotite, can be decomposed into a mixture of two sources, the first with a ratio of 1.9 times VSMOW assumed to be representative of the interior, and the second with a ratio of 5.2 times VSMOW, thought to be representative of the atmosphere[61].

A potential compounding factor is Carbon Dioxide. This will reduce the number of available sites for adsorption as all three gasses compete for a fixed amount of surface area. However, it has been found experimentally that water vapor is able to displace adsorbed CO_2 [62] and so this is expected to have little effect. The presence of CO_2 will also restrict the flow of molecules, however, it should be noted that the values of the diffusivity listed here are already much smaller than typical martian regolith diffusivities measured in the lab e.g. [40].

It is common practice that in the Knudsen regime, the molecular diffusivity of any additional interstitial gas can be ignored e.g., [38] since collisions occur much more frequently with the walls. This situation will change drastically once molecules have escaped the surface of the material, as molecular diffusion will dominate. Here advection of water vapor from the surface becomes important, as any back pressure will tend to restrict the diffusion rate by changing the pressure drop across the regolith. However, models have suggested that advection of the lighter H_2O and HDO in the heavier CO_2 on Mars may be more effective than sublimation into an atmosphere of pure water vapor [63] based on [64]. In fact, considering the model of [65], the typical advected flux at 203 K (H_2O pressure of ~0.5 Pa) is approximately 1.1×10^{16} molecules s^{-1} m^{-2}. This is comparable to the range of escape rates seen in the apparatus at this temperature of 1×10^{16} to 4×10^{18} molecules s^{-1} m^{-2}. As such, the results should be applicable to the Martian situation.

In the literature, present-day sublimation on Mars has only rarely been concerned with isotopic differences. Here, there are two schools of thought. The first of these is to consider ice on Mars as a continuously well mixed reservoir as one might consider the Earth's ocean [17] took this approach when they calculated total fractionation effects from surface to space [18] took a different approach, considering the opposite case in which no net fractionation occurred. Both of these models were attempts to use the enriched D/H ratio of Mars today to determine the size of the exchangeable water reservoir and both papers made important contributions [17] do an excellent job of characterizing the fractionation of the atmosphere to space, which is significant and has since been further refined e.g.[66-70] while [18] does an excellent job of quantifying the overall infalling water fluxes. However, neither model examined surface fractionation upon sublimation in detail. As such, our results provide the missing piece in the analysis.

Sublimation can itself bring about fractionation in surface reservoirs, as our results show, and much of the highly enriched material will remain close to the surface, likely adsorbed onto dust. As water moves on Mars mainly by vapor phase mechanisms, the enriched dust will gradually increase the amount of HDO present in the surface reservoir. Thus it is possible to imagine plausible scenarios in which Mars could appear to have a high-Deuterium surface and atmosphere without having lost as much water as would be required without the action of adsorbing dust. As such, direct sampling of the surface and subsurface water reservoirs is required to determine the bulk ratio, with coring of the polar cap being even more attractive as a means to deconvolve the isotopic effects of repeated volatile cycling over geologic time.

CONCLUSIONS

Sublimation experiments have been carried out to determine the effect of mixing dust with porous ice on the isotopic composition of the sublimate gas. Disaggregated samples, in particular, were produced with high porosities by flash freezing in liquid nitrogen and crushing to sand (mm) sized particles. When dust was incorporated in the mixture to be flash frozen, the D/H ratio of the sublimate gas was seen to decrease with time from the bulk ratio. The more dust was added to the mixture, the more pronounced was this effect.

Two possible mechanisms for producing this effect were identified, migration of material within the sample and adsorption on dust grains. Simple models for migration of HDO within the sample were prepared which included sublimation, condensation and vapor diffusion. However, these showed that it is very difficult to sequester isotopically heavy material at depth. This suggests that adsorption within a thickening dust mantle is a better mechanism for separating HDO and H_2O. This is bolstered by the fact that identical samples without dust did not fractionate. As well, without adsorption on dust the diffusivities observed within the dust mantle are too low to be credible.

A close examination of the binary adsorptive system showed that a difference in the adsorption coefficients of H_2O and HDO reproduces many of the attributes of the fractionating system. The model simulations are able to reproduce an increasing fractionation in the D/H ratio of the system with time and with increasing dust mixing ratio. The ratio of the adsorption coefficients can be determined from the fractionation to which the system asymptotes. For JSC Mars-1 near 200 K, an adsorption coefficient of HDO larger than that of H_2O by a factor of 0.6, i.e. $\alpha_{HDO} = \alpha_{H2O}/0.6$ as seen by [39] reproduces the observed trend in dust.

These results imply that dusty porous ices are more able to retain HDO than are clean porous ices. This has significance for the major reservoirs of sublimating dusty ices in the solar system; comets and the martian polar caps. In both cases, measurements of sublimated gas alone have been made and from these measurements a bulk ratio in the solid has been inferred. This work suggests that the estimates of the bulk ratios of these bodies need to be adjusted for the action of dust and must therefore take into account the history of volatile cycling on each body. As such, the dusty surfaces of comets may contain more deuterated material than their interiors making a determination of the primordial D/H ratio of cometary water from measurements made in cometary comae difficult at best. As for Mars, given plausible histories of water cycling on the planet, scenarios in which the atmospheric D/H ratio bears little resemblance to the bulk ratio of martian water are possible. As such, Mars may possess a greater fraction of its original water than previously considered possible.

In both cases, this work highlights the need for further experiments to determine the kinetic constants of adsorption and for ground truth. In particular, experiments which seek to determine the specific adsorption kinetic constants as a function of temperature down to 150 K would be highly valuable in further numerical modeling. In terms of ground truthing, only direct sampling of Martian and Cometary water will be able to reliably determine the D/H ratios of the different reservoirs. In fact, for Mars, coring of the polar ice cap may be necessary in order to determine typical values if the deposits of water ice are isotopically stratified from repeated cycling, as seems likely. Furthermore, measurements of D/H in comets that cover as much of the cometary orbit or atmosphere as possible could be used to shed light on the nature of water adsorbed on cometary regolith.

ACKNOWLEDGEMENTS

This work was partially supported by a scholarship provided by the Natural Sciences and Engineering Research Council of Canada (NSERC).

REFERENCES

1. Drake MJ: Origin of Water in the Terrestrial Planets.*Meteoritics and Planetary Science* 2005, 40:519.

2. Hunten DM: Atmospheric Evolution of the Terrestrial Planets. *Science* 1993, 259:5097.915-920

3. Owen T, Maillard JP, de Bergh C, Lutz BL: Deuterium on Mars: the Abundance of HDO and the value of D/H.*Science* 1988, 240(4860):1767-1770.

4. Krasnopolsky VA: High-resolution spectroscopy of Mars at 3.7 and 8 µm: A sensitive search of H_2O_2, H_2CO, HCl, and CH_4, and detection of HDO.*J Geophys Res* 1997, 102:6525-6534.

5. Hunten DM, Pepin RO, Walker JCG: Mass Fractionation in Hydrodynamic Escape. *Icarus* 1987, 69:532-549.

6. Geiss J, Gloeckler G: Composition of H, He, and Ne in the protosolar cloud. *Space Science Revs* 2003, 106:3-18.

7. Vanysek V: Isotopic Ratios in Comets. In *Comets in the Post-Halley Era*. Edited by Newburn RL, Neugerbauer M, Rahe J. Kluwer, Dordrecht, The Netherlands; 1991:299-311.

8. Eberhardt P, Reber M, Krankowsky D, Hodges RR: The D/H and O^{18}/O^{16} ratios in water from comet P/Halley. *Astronomy and Astrophysics* 1995, 302:301-316.

9. Bockelee-Morvan D, Gautier D, Lis DC, Young K, Keene J, Phillips T, Owen T, Crovisier J, Goldsmith PF, Bergin EA, Despois D, Wootten A: Deuterated water in comet C/1996 B2 (Hyakutake) and its implications for the origin of comets. *Icarus* 1998, 133:147-162.

10. Meier R, Owen TC: Cometary Deuterium. *Space Sci Rev* 1999, 90:33-43.

11. Gat JR: Oxygen and hydrogen isotopes in the hydrological cycle *Annu. Rev Earth Planet Sci* 1996, 24:225-262.

12. Matsuo S, Knuiyoshi H, Miyake Y: Vapor pressure of ice containing D_2O. *Science* 1964, 145(363):1454-1455.

13. Merlivat L, Nief G: Fractionnement isotopique lors des changements d'état solide-vapeur et liquide-vapeur de l'eau à des temperatures inférieures à 0°C. *Tellus* 1969, 19:122-127.

14. Daansgard W: Stable isotopes in precipitation. *Tellus* 1964, 16:436-468.

15. Van Hook WA: Vapor pressures of the isotopic waters and ices. *J of Phys Chem* 1967, 72:1234-44.

16. Van Hook WA: Vapor pressure isotope effect in aqueous systems. III. The Vapor Pressure of HOD (-60° to 200°). *J of Phys Chem* 1972, 76:3040-3043.

17. Yung YL, Wen J-S, Pinto JP, Pierce KK, Allen M: HDO in the Martian Atmosphere: Implications for the abundance of Crustal Water. *Icarus* 1988, 76:146-159. PubMed Abstract |

18. Carr M: D/H on Mars: Effects of Flood, Volcanism, Impacts and Polar Proceses. *Icarus* 1990, 87:210-227.

19. Brown RH, Lauretta DS, Schmidt B, Moores JE: Experimental and Theoretical Simulations of Ice Sublimation with Implications for the Chemical, Isotopic and Physical Evolution of Icy Objects. *Planetary and Space Science* 2012, 60(1):166-180. doi: 10.1016/j.pss.2011.07.023

20. Sears DWG, Kochan HW, Huebner WF: Laboratory simulation of the physical processes occurring on and near the surfaces of comet nuclei. *Meteoritics & Planetary Science* 1999, 34(4):497-525.

21. Huebner WF: The KOSI experiments. *Geophys Res L* 1991, 18(2):243-244.

22. Kochan H, *et al.*: Comet simulation experiments in the DFVLR space simulators. *Advances in Space Research* 1989, 9(3):113-122.

23. Gruen E, Kochan H, Seudenstickler KJ: Laboratory Simulation: A tool for comet research. *Geophysical Research Letters* 1991, 18:245-248.

24. Gruen E, Benkehoff J, Heidrich R, Hesselbarth R, Kohl H, Kuhrt E: Energy Balance of the KOSI-4 experiment. *Geophysical Research Letters* 1991, 18:253-256.

25. Gruen E, Bar-Nun A, Benkhoff J, Bischoff A, Dueren H, Hellmann H, Hesselbarth P, Keller HU, Klinger J: Laboratory simulation of comet processes: Results from the first KOSI experiments. In *In Comets in the Post-Halley Era.* Edited by Newburn RL, Neugebauer M, Rahe J. Dordrecht, The Netherlands: Kluwer; 1991:277-297.

26. Gruen E, Benkhoff J, Gebbhard J: Past, present and future KOSI comet simulation experiments. *Ann Geophysicae* 1992, 10:190-197.

27. Gruen E, Gebhard J, Bar-Nun A, Benkhoff J, Dueren H, Eich G, Hische R, Huebner WF, Keller HU, Klees G: Development of a dust mantle on the surface of an isolated ice-dust mixture: Results from the KOSI-9 experiment. *J Geophys Res* 1993, 98:15,091-15,104.

28. Seidensticker KJ, Kochan H, Mohlmann D: The DLR small simulation chamber: A tool for cometary research in the lab. *Advances in Space Research* 1995, 15(1):29-34.

29. Roessler K, Eich G, Klinger E, Trimborn P: Changes of Natural isotopic abundances in the KOSI comet simulation experiments. *Ann Geophysicae* 1992, 10:232-234.

30. Allen CC, Jager KM, Morris RV, Lindstrom DJ, Lindstrom MM, Lockwood JP: JSC Mars-1: A Martian soil simulant. *Proc Conf Amer Soc Civil Engineering* 1998, 469:-476. *Space 98*

31. Lemmon MT, *et al.*: Atmospheric imaging results from the Mars Exploration Rovers: Spirit and Opportunity. *Science* 2004, 306(5702):1753-1756.

32. Tomasko MG, Doose LR, Lemmon M, Smith PH, Wegryn E: Properties of dust in the martian atmosphere from the Imager for Mars Pathfinder. *J Geophys Res* 1999, 104(E4):8987-9007.

33. Pollack JB, Ockert-Bell ME, Shepard MK: Viking Lander image analysis of Martian atmospheric dust. *J Geophys Res* 1995, 100(E3):5235-5250.

34. Satoh H, Furukawa Y, Tsukamoto K: Isotope Segregation during Ice crystallization Process. *Nippon Kessho Seicho Gakkaishi* 2004, 31(3):284.

35. Lehmann M, Siegenthaler U: Equilibrium Oxygen and Hydrogen Isotope Fractionation between Ice and water. *Journal of Glaciology* 1991, 37(125):23-26.

36. Ibadinov KhI, Rahmonon AA, ASh Bjasso: Laboratory simulation of comet structures. In*Comets in the Post-Halley Era*. Edited by Newburn RL, Neugerbauer M, Rahe J. Dordrecht: Kluwer; 1991:299-311.

37. Wallace D, Sagan C: Evaporation of ice in planetary atmospheres: Ice covered rivers on Mars. *Icarus* 1979, 39:385-400.

38. Clifford SM, Hillel D: The stability of ground ice in the equatorial region of Mars. *J Geophys Res* 1983, 88:2456-2474.

39. Moores JE, Smith PH, Boynton WV: Adsorptive Fractionation of HDO on JSC MARS-1 during Sublimation with Implications for the Regolith of Mars. *Icarus* 2011, 211(2):1129-1149. doi:Doi: 10.1016/j.icarus.2010.10.020

40. Chevrier V, Ostrowski DR, Sears DWG: Experimental study of the sublimation of ice through and unconsolidatyed clay layer: Implications for the stability of ice on Mars and the possible diurnal variations in atmospheric water. *Icarus* 2008, 196:459-476.

41. Beck P, Pommerol A, Schmidt B, Brissaud O: Kinetics of water adsorption on minerals and the breathing of the Martian regolith. *J Geophys Res* 2010, 115:E10011. doi: 10.14029/2009JE003539

42. Hesselbarth P, Kankowsky D, Lammerzahl P, Mauersberger K, Winkler A, Hsiung P, Rossler K:Gas release from ice/dust mixtures. *Geophys Res L* 1991, 18:269-272.

43. Jänchen J, Morris RV, Bish DL, Janssen M, Hellwig U: The H_2O and CO_2 adsorption properties of phyllosilicate-poor palagonitic dust and smectites under martian environmental conditions. *Icarus* 2009, 200:463-467. doi:10.1016/j.icarus.2008.12.006

44. Pommerol A, Schmitt B, Beck P, Brissaud O: Water sorption on martian regolith analogs: Thermodynamics and near-infrared reflectance spectroscopy. *Icarus* 204(1):114-136. doi: 10.1016/j.icarus.2009.06.013

45. Koehler BG, Mak CH, Arthur DA, Coon PA, George SM: Desorption kinetics of hydrogen and deuterium from Si(111) 7 × 7 studied using laser-induced thermal desorption.*J Chem Phys* 1988, 89:1709. doi:10.1063/1.455117.

46. Hoogers G, Lesiak-Orowska B, King DA: Diffusion on a stepped surface: H and D on Rh332. *Surface Science* 1995, 327:47. doi:10.1016/0039-6028(94)00825-6 DOI:dx.doi.org

47. Lipshtat A, Biham O, Herbst E: Enhanced production of HD and D_2 molecules on small dust grains in diffuse clouds. *Monthly Notices of the Royal Astronomical Society* 2004, 348(3):1055-1064. doi:doi: 10.1111/j.1365-2966.2004.07437.x

48. Criss RE: *Principle of Stable Isotope Distribution*. New York, NY: Oxford University Press; 1999.

49. Brownlee D, *et al.*: Comet 81P/Wild 2 Under a Microscope. *Science* 2006, 314(5806):1711-1716. doi:DOI: 10.1126/science.1135840

50. Burnett DS: NASA Returns rocks from a comet. *Science* 2006, 314(5806):1709-1710. doi:DOI: 10.1126/science.1137084 m.

51. Jorda L, Lamy P, Faury G, Keller HU, Hviid S, Kuppers MM, Koschny D, Lecacheux J, Gutierrez P, Lara LM: Properties of the dust cloud caused by the Deep Impact experiment. *Icarus* 2006, 187:208-219.

52. Clifford SM, *et al.*: The State and Future of Mars Polar Science and Exploration. *Icarus* 2000, 144:210-242.

53. Langevin Y, Poulet F, Bibring J-P, Schmitt B, Douté S, Gondet B: Summer Evolution of the North Polar Cap of Mars as Observed by OMEGA/Mars Express. *Science* 2006, 307(5715):1581-1584.

54. Rickman H: The nucleus of comet Halley: Surface structure, mean density, gas and dust production. *Advances in Space Research* 1989, 9(3):59-71.

55. Hartogh P, Lis DC, Brokelée-Morvan D, de Val-Borro M, Biver N, Kuppers M, Emprechtinger M, Bergin EA, Crovisier J, Rengel M, Moreno R, Szutowicz S, Blake GA: Ocean-like water in the Jupiter-family comet 103P/Hartley 2. *Nature* 2011, 478(7368):218-220. doi:10.1038/nature10519

56. Podolak M, Mekler Y, Prialnik D: Is the D/H ratio of the Comet Coma equal to the D/H ratio in the Comet Nucleus? *Icarus* 2002, 160:208-211..

57. Hutsemekers D, Manfroid J, Zucconi JM, Arpigny C: The (OH)-O-16/(OH)-O-18 and OD/OH isotope ratios in comet

C/2002 T7 (LINEAR). *Astronomy and Astrophysics* 2008, 490:L31-L34.

58. Yung YL, Kass DM: Deuteronomy?: A Puzzle of Deuterium and Oxygen on Mars.*Science* 1998, 280(5369):1545-1546. PubMed Abstract |

59. Laskar J, Levrard B, Mustard JF: Orbital forcing of the martian polar layered deposits. *Nature* 2002, 419:375-377.

60. Leshin LA, Hutcheon ID, Epstein S, Stolper EM: Water on Mars: Clues from Deuterium/Hydrogen and Water Contents of Hydrous Phases in SNC Meteorites. *Science* 1994. 265 n°5168 86-90

61. Leshin LA: Insights into martian water reservoirs from analyses of martian meteorite QUE94201. *Geophys Res L* 2000, 27:14. 2017-2020

62. Zent AP, Quinn RC: Simultaneous adsorption of CO_2 and H_2O under Mars-like conditions and application to the evolution of the Martian climate. *J Geophys Res* 1995, 100(E3):5341-5349.

63. MacClune KL, Fountain AG, Kargel JS, MacAyeal DR: Glaciers of the McMurdo dry valleys: Terrestrial analog for Martian polar sublimation. *J Geophys Res* 2003, 108(E4):5031. doi:doi: 10.1029/2002JE001878

64. Ingersol AP: Mars: occurrence of liquid water. In *Planetary Atmospheres*. Edited by Sagan et. Al. Springer Verlag New York; 1971:247-250.

65. Toon OB, Pollack JB, Ward W, Burns JA, Bilski K: The astronomical theory of climactic change on Mars.*Icarus* 1980, 44:552-607.

66. Fouchet T, Lellouch E: Vapor Pressure Isotope Fractionation Effects in Planetary Atmospheres: Applications to Deuterium. *Icarus* 2000, 144:114-123.

67. Bertaux JL, Montmessin F: Isotopic fractionation through water vapor condensation: The Deuteropause, a cold trap for deuterium in the atmosphere of Mars. *J Geophys Res* 2001, 106(E12):32,879-32,884.

68. Krasnopolsky VA: Mars' upper atmosphere and ionosphere at low, medium and high solar activities: Implications for evolution of water. *J Geophys Res* 2002, 107(5128):E12. doi:doi: 10.1029/2001JE001809

69.   Cheng B-M, Chew EP, Liu C-P, Bahou M, Lee YP, Yung YL, Gerstell MF: Photo-induced fractionation of water isotopomers in the martian atmosphere. *Geophys Res L* 1999, 26(24):3657-3660.

70.   Miller CE, Yung YL: Photo-induced isotopic fractionation. *J Geophys Res* 2000, 105(D23):29,039-29,051.

Ground Water Contamination with Fluoride and Potential Fluoride Removal Technologies for East and Southern Africa

Bernard Thole[1]

[1]Physics and Biochemical Sciences Department, Polytechnic, Blantyre, University of Malawi, Malawi

INTRODUCTION

Ground water is main source of water supply in most rural communities in Africa. It has good microbiological and biological properties in general as such requires minimal treatment. Unfortunately groundwater is sometimes contaminated with naturally occurring chemicals. One such naturally occurring toxicant is fluoride. In some parts of Africa ground water contains high fluoride levels beyond the recommended World Health Organisation upper limit of 1.5 mg/l. It is reported that the East African Rift Valley is a high fluoride area. This region extends from Jordan valley down through Sudan, Ethiopia, Uganda, Kenya and Tanzania. High fluoride levels have also been reported in Malawi and The Republic of South Africa. In Kenya high fluoride levels in ground water beyond 5 mg/l and beyond 8 mg/l were reported in 20% and 30% respectively of 1000 samples taken nationally. A survey of fluoride in ground water in Tanzania showed that 30% of the waters used for drinking exceeded 1.5 mg/l fluoride. In Malawi and the Republic of South Africa fluoride levels beyond 1.5 mg/l and occurrence of dental fluorosis have also been reported. Proxy indicators of high fluoride levels in ground water are high pH, pH beyond 7, and high sodium and bicarbonate concentrations in the water. High fluoride waters often have low calcium and magnesium concentrations as such are fairly soft. There are some exceptions of fluoride occurrence that may not adhere to these proxy indicators.

The beneficial effects of ingesting fluoride to human health are limited to fluoride levels approaching 1.0 mg/l in potable water. It is reported that drinking of water with such levels of fluoride improves skeletal and dental health. Ingestion of water with fluoride levels beyond 1.5 mg/l has negative health impacts. Amounts in potable water between 1.5 and 3.0 mg/l will cause browning and mottling of teeth referred to as dental fluorosis. This is the onset of fluorosis that makes the teeth very hard and brittle. Concentrations between 4 to 8 mg/l result in skeletal fluorosis and crippling fluorosis ensues when water of greater than 10 mg/l fluoride is ingested for a prolonged period of time. Skeletal fluorosis is characterized by bone malformation resulting in movement difficulties while crippling fluorosis is characterized by weakening of the bones, and bone junctions growing together causing immobility. Excessive fluoride ingestion has other health effects reported in literature, among which are muscle fibre degeneration, low haemoglobin levels, red blood cell deformities, excessive thirst, headache, skin rashes, depression, gastrointestinal problems, urinary tract malfunction, nausea, abdominal pains, tingling sensation in fingers and toes, reduced immunity and neurological manifestations similar to pathological changes that occur in Alzheimer's disease patients. These effects have received less attention compared to dental and skeletal fluorosis that are typical in high fluoride areas.

Ingestion of fluoride through food and air is relatively small compared to fluoride ingestion through water. Attention has thus been drawn to controlling fluoride concentrations in water supplied for drinking. The World Health Organisation recommends that in mitigating for fluorosis in endemic areas the approach should be hierarchical in the following order; first to identify alternative source of potable water with low fluoride content, secondly to dilute high fluoride water with low fluoride water to attain a mass balance of within 1.5 mg/l, thirdly to use high calcium, magnesium and vitamin c diets and finally, when all these may not be feasible; to remove fluoride from water to meet the required level of 1.5 mg/l. Water defluoridation, the removal of fluoride from water, has been studied widely in time, space and materials. This is because the other lines of interventions are often not plausible in high fluoride rural areas where natural sources of water are used and income levels are humble. Wide research has resulted in an a lot of data and information on water defluoridation that may be employed in deciding for fluoride removal techniques at household, communal, municipal or regional level. However this information is oftentimes in different source materials and in different formats.

This chapter aims at enhancing progress towards access to safe drinking water through consolidating knowledge in groundwater fluoride occurrence, effects of fluoride on human health, and technologies available for water defluoridation in East and Southern Africa. Specifically the chapter will; provide information on fluoride occurrence in Eastern and Southern Africa and respective health effects to guide choices and decisions in water supply and treatment at municipal, regional and national level; consolidate research findings in water fluoride and defluoridation science for scientists and non-scientists that will assist in choices of water defluoridation technologies at home or local community and; exemplify research in special water treatment technologies through water defluoridation science for students in water resource sciences. To meet these objectives the chapter is outlined as follows;

FLUORIDE AND HUMAN HEALTH

The beneficial and harmful effects of fluoride ingestion are separated by a very narrow margin. Fluoride ingestion through potable water with concentrations about 1.0 mg/l is known to strengthen teeth and the skeleton; however water concentrations beyond 1.0 mg/l are undesirable because prolonged consumption of such water causes fluorosis. Dental fluorosis is caused by prolonged consumption of water with fluoride concentrations between 1.5 and 4.0 mg/l. This is characterised by browning and mottling of teeth. Prolonged drinking of water with concentrations of fluoride between 4.0 and 10 mg/l causes skeletal fluorosis and when water of concentrations beyond 10.0 mg/l is taken for a long time crippling fluorosis may ensue [1]. Skeletal fluorosis is characterised by weakening of bones and malformation of the skeleton. Symptoms of crippling fluorosis are the growing together of bone junctions causing immobility. The science behind the beneficial and harmful effects of fluoride on the skeletal structure is based on the possible ion exchange reactions between hydroxide and fluoride ions in the calcium hydroxy-phosphate, the main skeletal structure compositional material. The replacement of hydroxide ions with fluoride ions, Equation 1, results in a more acid resistant structure, fluoroapatite.

$$Ca_5(PO_4)_3OH + F^- \rightarrow Ca_5(PO_4)_3F + OH^-$$

(1)

Fluoroapatite being more resistant to acid attack compared to hydroxyapatite offers a protective layer to the tooth enamel against acids from foods. This prevents dental caries. Excessive fluoride intake however may enhance the reaction to go beyond replacement of hydroxide, Equation 2.

$$Ca_5(PO_4)_3F + 9F^- \rightarrow Ca_5F_{10} + 3PO_4^{3-}$$

(2)

In Equation 2 ion exchange occurs between phosphate and fluoride ions. The resultant compound, calcium decafluoride, is a very hard and brittle material not appropriately suited for the functions of the skeletal structure [2]. Other complications associated with excessive consumption of fluoride are muscle degeneration, low heamoglobin content, deformities of red blood cells, skin rashes, depression, abdominal pains, urinary tract malfunction, reduced immunity, tingling sensation in fingers and toes, excessive thirst, and, neurological manifestations similar to pathological changes that occur in Alzhemer's disease patients [3]. Dental and skeletal fluorosis has however attracted greater attention as compared to the other effects of fluoride because of its obvious manifestations. In Malawi for instance, high correlation was obtained between levels of fluoride in drinking water with occurrence of dental fluorosis in primary school pupils. High correlation between fluoride levels in groundwater and occurrence of dental fluorosis was also obtained in a number of districts in the country. Significant correlation ($r^2 = 0.77$) between levels of fluoride in borehole water and manifestation of dental fluorosis in primary school pupils of Liwonde, a township in Southern Malawi. A similar picture emerges in Nathenje, a township in Central Malawi, where 68.5 % of school going children in high fluoride areas showed signs of dental fluorosis [4, 5].

Incidences of fluorosis have been reported in the Republic of South Africa in high fluoride areas. Research has shown that 803 areas are fluorosis endemic in South Africa. These areas include locations in Western and Karoo Regions of Cape Province, the North Western, Northern, Eastern and Western areas of Transvaal, Western and Central Free State. A study on dental fluorosis occurrence among children revealed that even at low concentrations of fluoride in potable water dental fluorosis ensues. In sub optimal fluoride areas (0.4 – 0.6 mg/l) dental fluorosis was evidenced in about 19 % of children [3]. Results obtained, illustrate that there may be no universal safe levels of fluoride in drinking water [6]. Significant dental fluorosis incidences were noted in low, medium and high fluoride areas, Table 1.

Table 1: Fluoride and dental fluorosis occurrence in three locations of South Africa; [3, 6]

Location	Fluoride concentration in potable water	% with Dean's index score greater than 2.
Sanddrif	0.19	47
Kuboes	0.48	50
Leeu Gamka	3.0	95

Much higher fluoride levels occur in Tanzania and high dental and skeletal fluorosis have been reported in the Kilimanjaro region. Among 119 children aged between 9 and 13 severe dental fluorosis was evident in 87.4 % of the children at Maji ya chai in Meru, Tanzania. These children had lived all their lives within this area and drank water from a river with fluoride levels of 18.6 mg/l. Very high occurrence of dental fluorosis was also reported among adults in Arusha, 83 %, and in Moshi, 95 %, in Tanzania. Regions most affected in Tanzania are Arusha, Moshi, Singida and Shinyanga. In Arusha skeletal fluorosis has been observed [3, 7].

Dental fluorosis in Kenya has been reported as having a prevalence rate of up to 39.6 % in three racial groups. Table 2 illustrates the distribution of signs of dental fluorosis classified using the extent of affliction;

Table 2: Reported prevalence of dental fluorosis in Kenya; [8]

Distribution of dental fluorosis (%) ⇓	Race ⇒	African	Asian	European
Sample size		3,014	626	922
Normal		46.4	30.4	61.3
Questionable		15.7	11.7	15.7
Very mild		17.9	17.5	13.4
Mild		10.8	28.6	6.4
Moderate		5.5	5.7	2.3
Severe		3.7	6.1	0.9
Prevalence (%)		37.9	57.9	23.0

Areas most affected in Kenya are the Northern Frontier (Turkana), Northe-West Kenya, Southern Rift Valley, Central and Eastern Regions. Surveys found that 67 % of Asian, 47 % of African and 30 % of European school children showed signs of dental fluorosis of varying degree. The high prevalence in Asian population was speculated to relate to their vegetarian diet [3]. This research carried out in Kenya employed Dean's index that is decribed below.

Table 3: Dean's Index; [9]

Class	Description
1. Normal	Complete absence of any white flecks or white spots
2. Questionable	A few white flecks to occasional white spots.
3. Very mild	Less than 25 per cent of the tooth surfaces covered by small white opaque areas

4. Mild	Fifty per cent of the tooth surfaces covered by white opaque areas.
5. Moderate	Nearly all the tooth surfaces are involved, with minute pitting and brown or yellowish stains.
6. Severe	Smoky white appearance of all the teeth with hypoplasia, chipping and large brown stains, which vary from chocolate brown to black. There is discreet and confluent pitting, often accompanied by attrition.

There are other indices that are often employed. These are; DDE (Developmental Defects of Enamel) index developed by Federation Dentaire Internationale in 1992; Thylstrup Fejerskov (TF) index by Thylstrup and Fejerskov (1978) and Tooth Surface Index of Fluorosis (TSIF) developed by Horowitz et al., in 1984 [3].

Table 4: USPHS Fluoride recommendations in drinking water

Annual average of maximum daily air temperature (°C)	Recommended fluoride concentration (mg/l)			Maximum allowable fluoride concentration (mg/l)
	Lower	Optimum	Upper	
10 – 12	0.9	1.2	1.7	2.4
12.1 – 14.6	0.8	1.1	1.5	2.2
14.7 – 17.7	0.8	1.0	1.3	2.0
17.8 – 21.4	0.7	0.9	1.2	1.8
21.5 – 26.2	0.7	0.8	1.0	1.6
26.3 – 32.5	0.6	0.7	0.8	1.4

Many parts of some countries in Africa are also affected by fluorosis. These include some areas in Sudan, Uganda, Ethiopia, Senegal and Niger. Fluoride ingestion is highly linked to drinking water because the contributions from other sources, such as food and air is minimal. The United States Public Health Service [10] set some guidelines for lower, optimal, upper and maximum allowable fluoride concentrations in drinking water with respect to average air temperature, Table 3:

FLUORIDE OCCURRENCE IN THE WORLD

Geogenic occurrence of fluoride is often linked to volcanic activity, fumaric gases and presence of thermal waters. Proxy indicators of high fluoride levels in groundwater are; low levels of calcium and magnesium, high levels of sodium and bicarbonate ions, and high pH above 7. There are however some exceptions to these generic conditions [11]. High fluoride ground waters are typically of sodium chloride, or sodium chloride bicarbonate type characterised by high pH. Areas with high fluoride in ground water include fluoride beds encompassing parts of Iraq, Iran, Syria, Turkey, Algeria and Morocco, and the East African rift system extending from Jordan valley down through Sudan, Ethiopia, Uganda, Kenya and Tanzania. There are high fluoride areas in other parts of the world, Figures 1 to 6show high fluoride areas of the world.

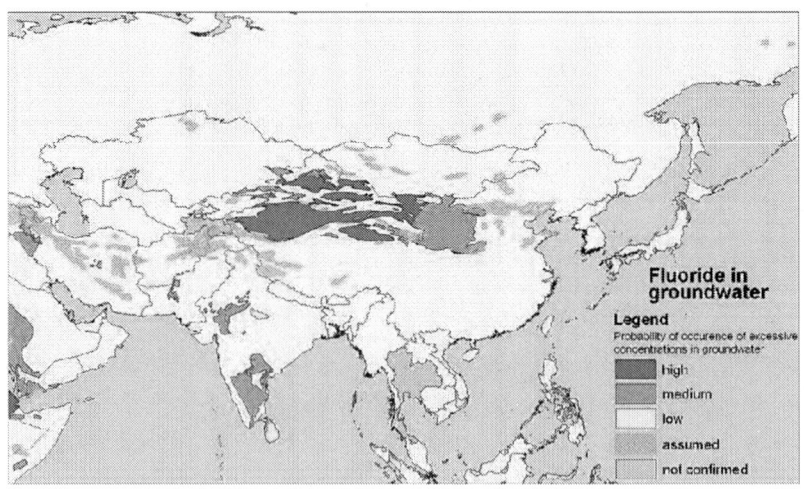

Figure 1: Fluoride occurrence in groundwater in Asia [11].

Fluoride is found in a wide variety of minerals that include fluorspar (CaF_2), cryolite (Na_3AlF_6), apatite ($Ca_5(PO_4)_3F$) and hornblende [$(ca,Na)_2(Mg,F,Al)_5(Si,Al)_8O_{22}(OH)_2$]. The average crustal abundance is known to be about 300 mg/kg representing between 0.06 to 0.09 % by weight of the earth crust. The presence of fluoride in ground water results from dissolution of fluoride bearing minerals where the water is in contact with a fluoritic bed.

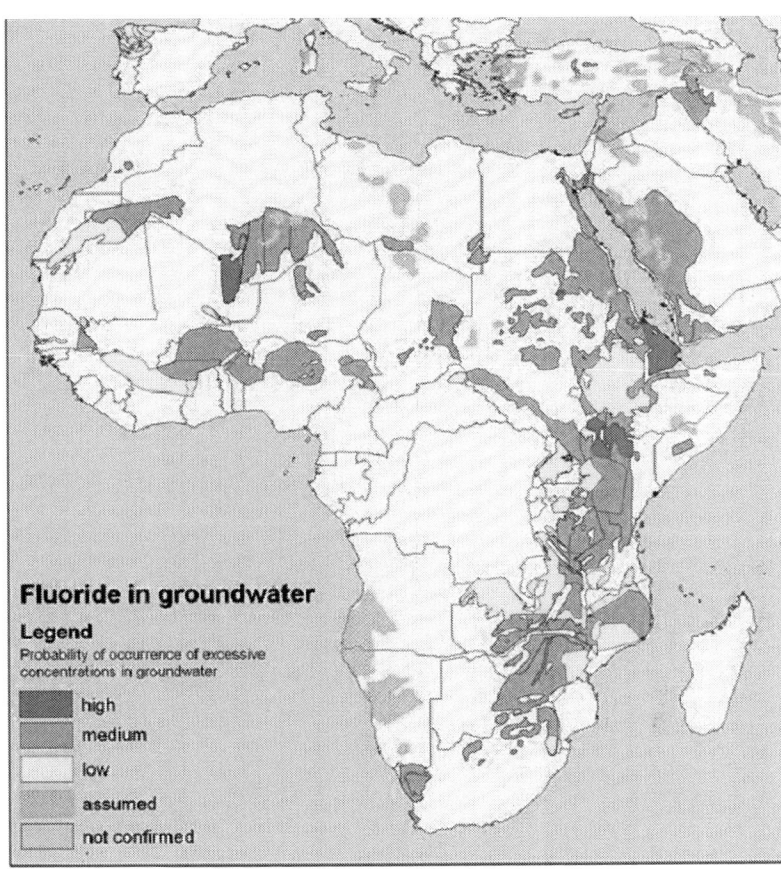

Figure 2: Fluoride occurrence in groundwater in Africa [11].

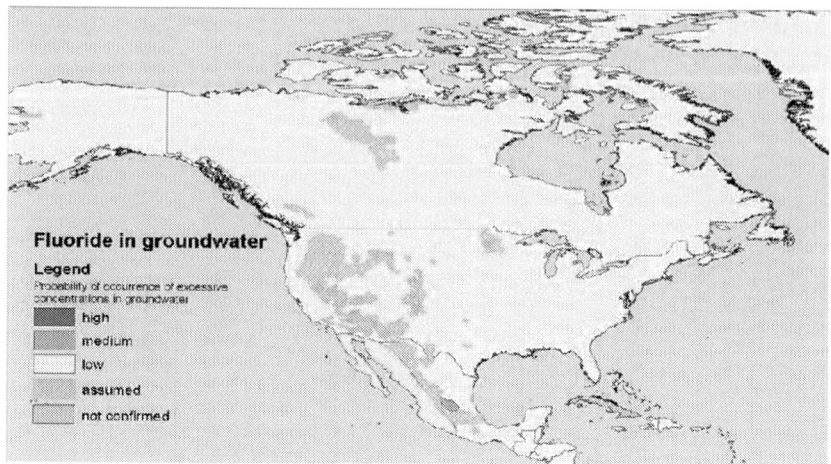

Figure 3: Fluoride occurrence in groundwater in North and Central America [11].

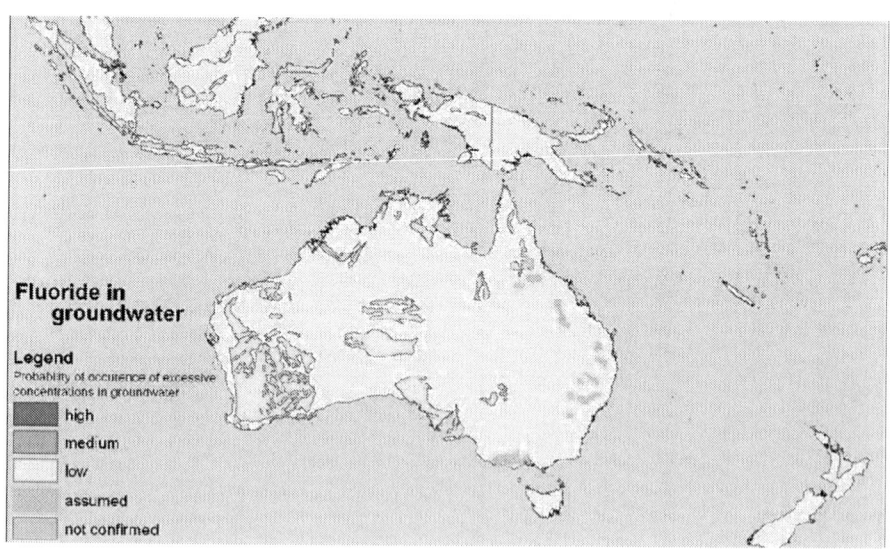

Figure 4: Fluoride occurrence in groundwater in Oceania [11].

Figure 5: Fluoride occurrence in groundwater in South America [11].

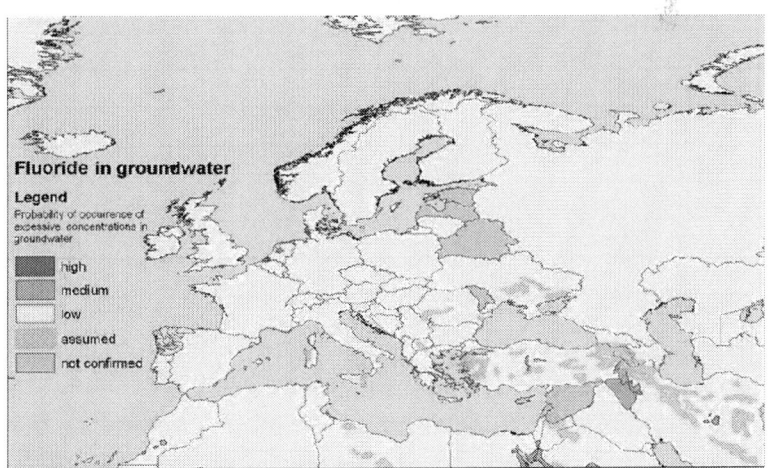

Figure 6: Fluoride occurrence in groundwater in Europe [11].

FLUORIDE OCCURRENCE IN EAST AND SOUTHERN AFRICA

High fluoride levels occur in ground waters in some parts of Kenya, Tanzania, Malawi and The Republic of South Africa; however the East African countries have higher levels compared to the Southern Countries. In Tanzania for example, fluoride concentrations in ground water of up to 40 mg/l have been reported, see Table 5. Some lakes in East Africa have extremely high fluoride concentrations, an occurrence not typical in surface waters. Lake Elmentaita and Lake Nakuru of Kenya have fluoride concentrations of 1,640 mg l–1 and 2,800 mg l–1 respectively [12]. The Tanzanian Lake Momella is reported to have a fluoride concentration of 690 mg/l [3].

Table 5: Fluoride concentrations in some ground water sources in Northern Tanzania

Location	Average fluoride level in ground water
Arusha Maji ya Chai, Arumeru District	20.0
Lemongo spring 10.5	10.5
Kikati B/H 113/79 11.0	11.0
Masai Furrow- Tingatinga 32.0	32.0
B/H 186/81 - Hanang 46.0	46.0
Singida S/W 8/78 - Ngorongoro 11.6	11.6
Senene 10.5	10.5
Well camp Doromoni 21.3	21.3
Fish camp Migilango village 12.5	12.5
Hot spring - Manyoni 10.5	10.5
Shinyanga S/W Mkokolo 17.0	17.0

[i] - Sourced from Ngurdoto Defluoridation Research Centre, Tanzania

A detailed survey of fluoride concentrations carried out in Kenya revealed that 20 % of the samples had fluoride levels greater than 5 mg/l and 12 % of the samples exceeded 8 mg/l. A total of 1000 samples were taken from different locations covering the whole country. The highest concentrations were reported in ground waters of the volcanic areas of the Nairobi, Rift Valley and Central Provinces where maximum groundwater fluoride concentrations were as high as 30–50 mg/l [12]. In Tanzania concentrations of up to 45 mg/l have been detected in the rift valley. The most affected areas of Tanzania are Mwanza, Mara, Shinyanga, Arusha, Kilimanjaro and Singida shown in shaded lines inFigure 7.

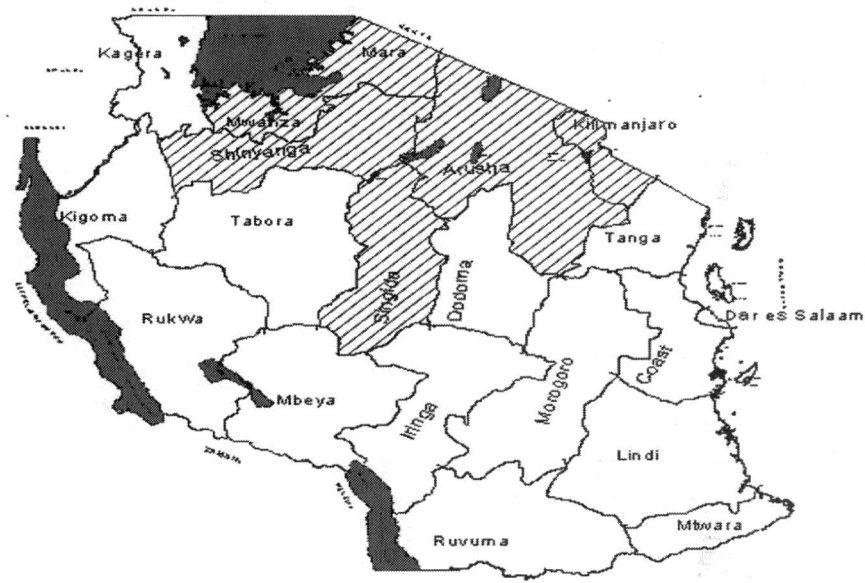

Figure 7: Map of Tanzania showing areas most affceted with fluoride (Sourced from Ngurdoto Defluoridation Research Station).

Fluoride occurrence in groundwater in Malawi has been better surveyed in the southern region [13, 14,15]. Some data extracted from research is depicted in Table 6.

Research in the Republic of South Africa has shown that underground mine waters may contain high fluoride levels beyond 3 mg/l. In one selected case fluoride levels of about 6 mg/l were identified in groundwater of Madibeng Local Municipality, North West Province of South Africa. West province is one of the areas in South Africa where fluorosis is typical. Areas affected include the North-West provinces, the Karoo, Limpopo and the Northern Cape. Cases like these have attracted research in water defluoridation such that evaluation of activated alumina as a defluoridating agent was carried out [17 - 19]. The researches demonstrated that the activated alumuna could be employed to treat underground mine water with initial fluoride levels as high as 8 mg/l. Two defluoridation plants were installed each with capacity of 500, 000 litres/day in the early 1980's in the Republic of South Africa.

Table 6: Some locations with high fluoride levels in groundwater reported in literature

Location	District	Fluoride level in ground-water (mg/l)	Reference
Bangula market	Nsanje	4.91 ± 0.03	[13]
Nsanje level crossing	Nsanje	7.25 ± 0.01	[16]
Tomali trading centre	Chikwawa	1.91 ± 0.00	[15]
Tomali dip tank	Chikwawa	1.93 ± 0.01	[15]
Mlangalanga Village, Malindi	Mangochi	2.60 ± 0.00	[13]
Mangochi hospital	Mangochi	2.45 ± 0.01	[13]
Nsauya 1	Mangochi	3.64 ± 0.01	[13]
Mbando village	Zomba	6.51 ± 0.01	[13]
Mtubwi	Machinga	7.51 ± 0.00	[14]
Mliwa village	Machinga	5.60 ± 0.00	[13]
Evangelical Baptist Church	Machinga	5.08 ± 0.01	[13]
Machinga hospital	Machinga	4.73 ± 0.01	[14]
Duwa village	Machinga	4.88 ± 0.00	[13]
Chedweka	Machinga	6.47 ± 0.02	[13]
Mazengera	Lilongwe	7.00 ± 0.01	[5]
Nkhotakota boma	Nkhotakota	9.60 ± 0.02	[16]
Songwe	Karonga	8.00 ± 0.01	[16]

RESEARCH ON FLUOROSIS AND WATER DEFLUORIDATION

Research on removal of fluoride from drinking water has employed very many materials world over. The main principles however remain adsorption, ion exchange, precipitation, coagulation, membrane processes, distillation and electrolysis.

Adsorption and Ion Exchange

Adsorption involves passage of water through a contact bed where fluoride is removed by ion exchange or surface chemical reaction with the solid bed matrix. After a period of operation, a saturated column must be refilled or regenerated. The different adsorbents used for fluoride removal include activated alumina, carbon, bone charcoal, activated alumina coated silica gel, calcite, activated saw dust, magnesia, serpentine, tricalcium phosphate, activated soil sorbents, carbion, defluoron, and other synthetic ion exchange resins [20 – 24]. Most widely used adsorbents are activated alumina and activated carbon. Activated alumina was first proposed and researched for defluoridation around 1930. It is basically highly porous aluminium oxide with large surface area. The discontinuous cationic lattice of alumina gives it localized areas of positive charge. This renders alumina a good adsorbent for many anionic species, however its greater preference for fluoride compared to other ions has led to its wide use in defluoridation [25]. Defluoridation capacity of activated alumina decreases with increase in hardness. High fluoride concentrations increase the solubility of alumina due to the formation of monomeric aluminium fluoride and aluminium hydroxyl fluoride complexes. It is established that defluoridation with alumina is optimal at pH 5 – 6. The activated alumina does not shrink, swell, soften nor disintegrate when immersed in water but dissolves at pH less than 5.0. At pH greater than 7.0 silicate and hydroxide compete strongly with fluoride for adsorption/exchange sites on alumina resulting in lower fluoride sorption. Chloride does not interfere with sorption of fluoride on activated alumina [23]. The use of activated alumina is highly selective towards fluoride but the pH specificity, low capacity and low material integrity in acidic medium are some of the limitations of this process [20].

It is common to first treat alumina with hydrochloric acid to make it acidic. However this treatment is often carried out in acidic medium with pH between 5 and 6 to avoid excessive dissolution of the alumina that occurs below pH 5. The chloride ions on the acidic alumina are replaced by fluorides when the alumina is in contact with fluoride ions. Equations 3 and 4 below illustrate the activation and ion exchange processes respectively.

$$Al_2O_3{}^\Phi H_2O + HCl \rightarrow Al_2O_3{}^\Phi HCl + H_2O \tag{3}$$

$$Al_2O_3{}^\Phi HCl + NaF \rightarrow Al_2O_3{}^\Phi HF + NaCl \tag{4}$$

$Al_2O_3{}^\Phi$ indicates activated alumina)

To regenerate the adsorbent a dilute solution of sodium hydroxide is mixed with the adsorbent to get a basic alumina, equation 5, followed by further treatment with acid, equation 6.

$$Al_2O_3{}^\Phi HF + 2NaOH \rightarrow Al_2O_3{}^\Phi NaOH + NaF + H_2O \tag{5}$$

$$Al_2O_3{}^\Phi NaOH + 2HCl \rightarrow Al_2O_3{}^\Phi HCl + NaCl + H_2O \tag{6}$$

The regeneration yields sodium fluoride concentrated wastewater that requires disposal, another challenge in the alumina adsorption process. Adsorption with activated carbon is another efficient technique, however, high cost and challenges with the spent carbon limit its large scale application [26]. Granular activated carbon is often employed in adsorption columns because of its non-specific nature of adsorption on its surface [27]. Powdered activated carbon has also been employed successfully in water defluoridation despite that the process is highly pH dependent with optimum results below pH 3 [28]. This requires pH reduction during defluoridation and increasing pH artificially in the water after treatment which is a challenge in the employment of this process. Bone char, which is derived from animal bones charred at 500–600°C, has a rich surface of heterogeneous components, allowing physisorption, chemisorption or ion exchange to occur. Physisorption, also termed physical adsorption, is adsorption that involves van der Waals forces (intermolecular forces) and there are no significant changes in the electronic orbital patterns of the species involved. Chemisorption, on the other hand, involves valence forces of the same kind as those that result in the formation of chemical compounds. The combination of physisorption, chemisorption and ion exchange processes renders bone char a better sorbent, in terms of ion uptake capacity, among other carbon based adsorbents such as activated carbon and peat [29, 30]. Bone charcoal has been widely applied in water defluoridation (Castillo et al., 2007 [7, 31, 32]. It is typically a black, granular and porous material with about 57 to 80 percent calcium phosphate [

$Ca_3(PO_4)_2$], 6 to 10 percent calcium carbonate ($CaCO_3$) and 7 to 10 percent activated carbon. Principal reaction in defluoridation with bone charcoal is hydroxyl-fluoride exchange of apatite, equation 7 [3].

$$Ca_{10}(PO_4)_6(OH)_2 + 2F^- \rightarrow Ca_{10}(PO_4)_6F_2 + 2OH^-$$

(7)

The preparation of bone charcoal is crucial. Unless carried out properly, the bone charring process may result in a product of low defluoridation capacity and/or deterioration in aesthetic water quality. Water treated with poor bone charcoal may taste and smell like rotten meat and is aesthetically unacceptable [3, 7].

Precipitation – Coagulation

The Nalgonda technique is a widely known precipitation – coagulation defluoridation method. In this technique aluminium sulfate and lime are added periodically in batch to treat water. These are co-precipitation chemicals that behave as shown, equations 8 - 11, in fluoride removal.

$$Al_2(SO_4)_3 + 6H_2O \rightarrow 2Al(OH)_3 + 3SO_4^{2-} + 6H^+$$

(8)

$$Al(OH)_3 + F^- \rightarrow Al-F_{Complex} + undefined_{ppt}$$

(9)

$$CaO + H_2O \rightarrow Ca(OH)_2$$

(10)

$$3Ca(OH)_2 + 6H^+ \rightarrow 3Ca^{2+} + 6H_2O$$

(11)

Other co-precipitation chemicals such as polyaluminium chloride

(PAC), lime and similar compounds are also employed and are added daily to raw water in batches. Precipitation techniques produce a certain amount of sludge every day [3].

Calcium and phosphate compounds are an example of contact precipitation chemicals often added to the water upstream of a catalytic filter bed. In contact precipitation there is no sludge and no saturation of the bed, only the accumulation of the precipitate in the bed. Theoretically it is possible to precipitate fluoride as calcium fluoride or fluoroapatite in solutions containing calcium, phosphate and fluoride; however in practice it is kinetically impossible. The reaction kinetics are very slow. Precipitation is easily catalysed in a contact bed that acts as a filter for the precipitate. The reactions involve dissolution of calcium chloride and sodium dihydrogen phosphate, and consequent precipitation of calcium fluoride and fluoroapatite [33]. These are illustrated through equations 12 to15.

$$CaCl_{2(aq)} \rightarrow Ca^{2+}_{(aq)} + 2Cl^-_{(aq)} \tag{12}$$

$$NaH_2PO_{4(aq)} \rightarrow Na^+_{(aq)} + 2H^+_{(aq)} + PO_4^{2-}_{(aq)} \tag{13}$$

$$Ca^{2+}_{(aq)} + 2F^-_{(aq)} \rightarrow CaF_{2(s)} \tag{14}$$

$$^{2+}_{(aq)} + 6PO_4^{2-}_{(aq)} + F^-_{(aq)} \rightarrow Ca_{10}(PO_4 \tag{15}$$

A small saturated bone char contact bed is employed as a column. This column is supported by coarse grain charcoal or gravel. Figure 8 shows the Ngurdoto configuration [34].

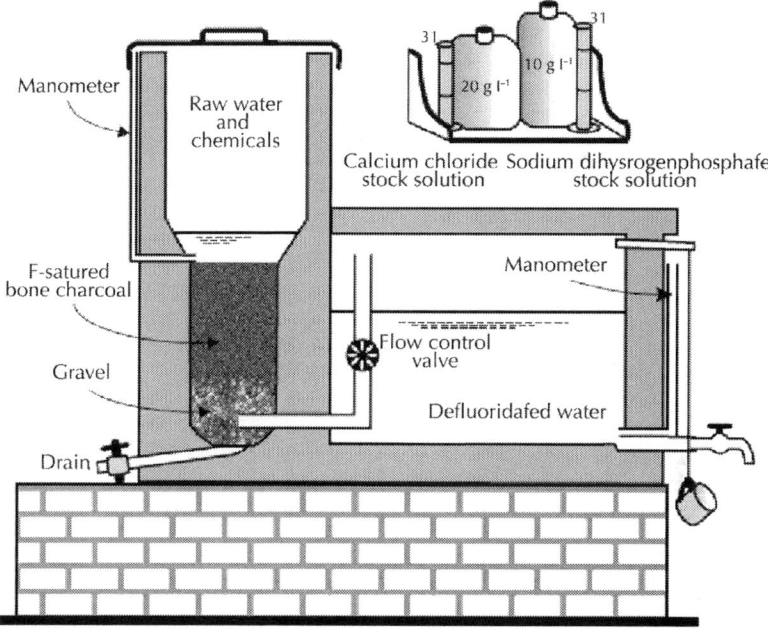

Figure 8: Contact precipitation of fluoride as invented in Ngurdoto.

It is recommended to prepare the chemicals monthly as stock solutions and employ them in aliquots. The two stock solutions should not be mixed before treatment in order to avoid the precipitation of calcium phosphate. Two special measuring cups may be used for volumetric portioning of the chemicals. It is advisable to check the bulk density as it may vary for different brands. The stock solutions, stored in Jerry cans, along with the respective chemical bags and the measuring cups and cylinders may be coloured red and green respectively in order to minimize the risk of exchange and so incorrect dosage [3]. The design criteria of the contact precipitation plants is simple however, the theoretical background is fairly complex and largely dependent on the reactions shown in equations 14 and 15. The extent to which each reaction occurs is not well understood. In calcium fluoride precipitation, the Ca/F weight ratio is about 1, equivalent to a CC/F ratio of about 4. In fluorapatite precipitation, the Ca/F ration is 11 and the PO_4/F ratio is 15, equivalent to a CC/F ratio of about 39 and a MSP/F ratio of about 23. This implies that the more fluoride precipitated as calcium fluoride, rather than as fluorapatite, the lower is the required dosage of chemicals. Calcium fluoride precipitation is probably more dominant with higher raw water fluoride concentration. Experience from operations of the contact precipitation in Tanzania, where the fluoride concentration averages 10 mg/l, has shown that the process functions effectively when the dosage ratios are 30 and 15 for CC and MSP respectively. This dosage would ensure at least 65 per cent precipitation of fluorapatite and a surplus of calcium for precipitation of the residual fluoride as calcium fluoride [33, 34].

Membrane Filtration

Membrane filtration processes are among advanced water treatment technologies that have been mainly employed in treatment of pure and ultra-pure water. The US EPA, 2003, recommended reverse osmosis, RO, as one of the best available defluoridation technologies [35]. Reverse osmosis and nano-filtration (NF) are the well-known membrane technologies that can remove a large spectrum of contaminants from water such as pathogens, turbidity, heavy metals, salinity, natural and synthetic organics, and hardness [36]. The two processes are highly effective in water defluoridation and produce high quality water that includes disinfection during water treatment. NF membranes operate at a lower pressure and have lower capacity as compared to RO membranes. Another membrane technology is electrodialysis. Use of electrodialysis plants in North Africa is employed in large scale water treatment of high fluoride brackish water for potable water supply [37]. Electrodialysis is similar to Reverse Osmosis, except it uses an applied direct current potential instead of pressure, to separate ionic contaminants from water. Water does not physically pass through the membrane in the electrodialysis process as such particulate matter is not removed. The ED membranes are therefore not technically considered filters. The water quality from electrodialysis treatment is comparable to RO, and may require post-treatment stabilization. The process tends to be most economical for source water with TDS levels in excess of 4,000 mg/L. It is established that RO and electrodialysis have very high defluoridation capacities (85 – 95 %) and function effectively in any pH range. However the water loss is high (20 - 30 % for electrodialysis, 40 – 60 % for RO), have high capital cost and are energy intensive [37]. Membrane technologies often require special equipment, electrical energy and specialized training for operators as such the capital and operation costs are high. Low applicability is therefore envisaged for rural sectors of the developing countries where energy and trained human resource are often deficient.

Emerging Technologies

Some emerging technologies employing precipitation, distillation and /or a combination of principles are; The Crystalactor®, Memstill® technology, The WaterPyramid® solution and The Solar Dew Collector system. The Crystalactor® was developed by DHV in the Netherlands [38]. It is a pellet reactor employing a fluidized bed. Water defluoridation occurs in the reactor accompanied by formation of calcium fluoride pellets of 1 mm diameter. The Crystalactor® employs contact precipitation and has the strengths that; the installation is compact, produces usable calcium fluoride pellets with high-purity, and the produced pellets have extremely low water content (5% to 10% moisture). It is estimated that this technology costs about a quarter of the conventional precipitation techniques. The technology is however suitable for treating high fluoride waters (>10 mg/l) and to attain concentrations below 1 mg/l a second treatment is often required. A membrane based distillation concept is also reported as developed by the Netherlands Organisation of Applied Scientific Research (TNO), the Memstill® technology, Figure 11. This technology advances ecology and economy of the existing technologies in brackish and sea water desalination. The technology also removes other anions such as fluoride and arsenic [39]. In the Memstill® technology cold feed water takes up heat in the condenser channel through condensation of water vapour, then a small amount of (waste) heat is added, and flows counter currently back via the membrane channel. This small added heat evaporates water through the membrane. The water is discharged as cold condensate. The cooled brine is disposed, or extra concentrated in a next module. The Memstill® technology can produce potable water at a cost well below that of existing technologies like reverse osmosis and distillation. It is expected that the Memstill® technology will also be developed for small scale applications using solar heat [39].

Principle of Memstill-process

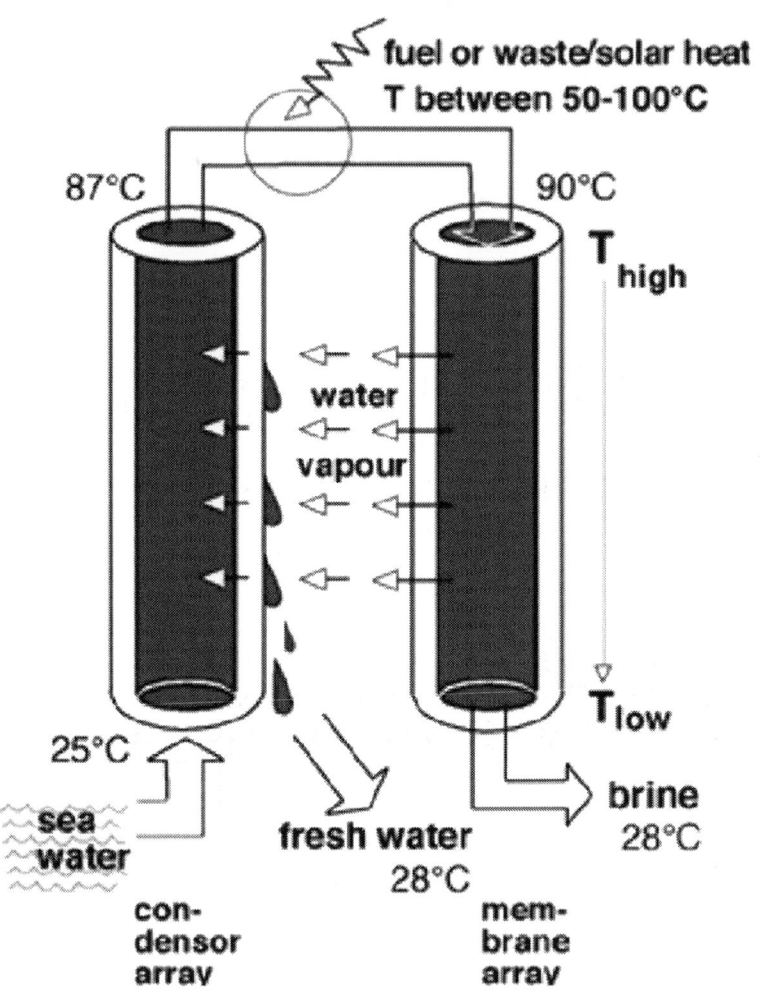

Figure 9: Memstill® technology [39].

The water pyramid, developed for rural tropical areas, employs solar energy to produce potable water from saline, brackish or polluted water, Figure 12 [40]. The technology also removes fluoride. A water pyramid with a total area of 600 m², placed under favourable tropical conditions, can produce about 1250 litres of fresh water a day. The rate of production is however dependent on local atmospheric conditions such as climate, temperature, cloud-cover and wind activity. Solar energy drives the desalination while energy required for pressuring the Water Pyramid® is obtained using solar cells combined with a battery backup system. A small generator may be required to cater for intermittent peak demands in electricity.

Figure 10: The Water Pyramid® [40].

Another technique similar to the Water Pyramid was developed by Solar Dew, Figure 13 [41]. This is a porous membrane that purifies water using solar energy. In this techniques water sweats through a membrane and evaporates on the membrane surface. This increases humidity in the evaporation chamber. As a result of temperature difference pure water condenses on the cooler surface of the system.

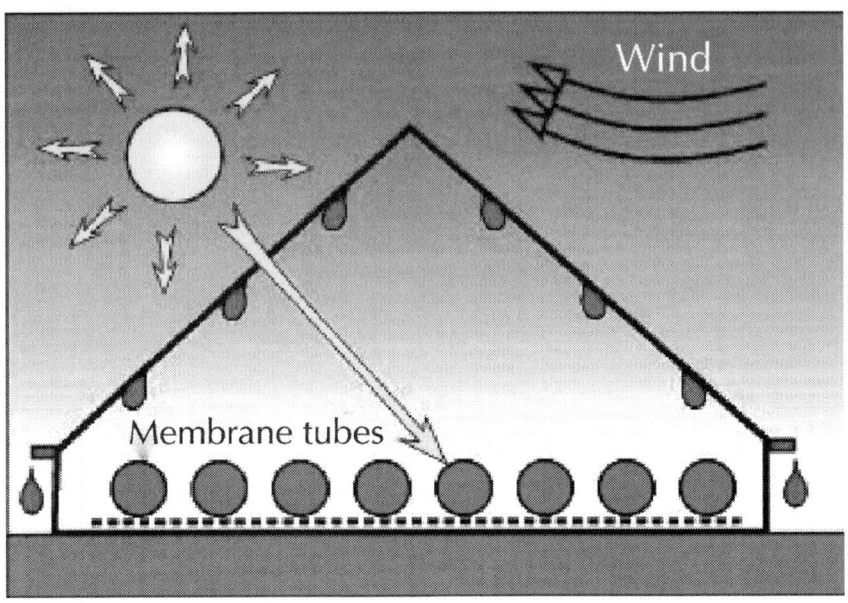

Figure 11: The Solar Dew Collector system [41].

Larsen and Pearce, 2002, proposed a defluoridation method in which fluoride containing water is boiled with brushite ($CaHPO_{4}·_{2}H_{2}O$) and calcite ($CaCO_{3}$). Good results were obtained on laboratory scale. Larsen and Pearce concluded that boiling brushite and calcite in fluoritic water yields fluoroapatite which results in defluoridation [42].

RESEARCH ON FLUOROSIS AND WA-TER DEFLUORIDATION IN EAST AND SOUTHERN AFRICA

Research was carried out in Tanzania at Ngurdoto Defluoridation Research Station employing different materials in batch and fixed bed defluoridation configurations. In Malawi Research was carried out at Chancellor College of the University of Malawi. Bone char, bauxite, gypsum, magnesite have been experimented on in defluoridation.

Water Defluoridation Research in Tanzania

Removal of fluoride from water has been researched on employing a number of materials among which are bauxite, gypsum, magnesite and their composite filters [2,46,47], Calcium chloride ($CaCl_2$) and Sodium dihydrogen phosphate ($NaH2PO_4$) as co-precipitation reagents in contact precipitation [33,34], cow bone char in batch and fixed bed [7, 32], fish bone [43], activated carbon and activated carbon loaded separately with alumina, magnesia and calcium [44], and, magnesite [45].

Research on defluoridation with bauxite, gypsum and magnesite focused on developing a hybrid technology with the three materials to reduce negative impacts on water quality that are encountered when each of the materials i.e. bauxite, gypsum and magnesite, is employed alone. The composites tested were mixtures of bauxite, gypsum and magnesite in respective mass ratios of 1:2:3, and such combinations with the order of the ratio numbers varied, giving a total of six compositions. Different calcine temperatures ranging from 150 to 500 °C were also tested and performance of batch and fixed bed configurations were compared. The research results in summary showed that the fixed bed configuration of the 3:2:1 (mass ratio of bauxite, gypsum, magnesite respectively) obtained water of optimum quality when calcined at 200 °C. The water quality parameters included in this research were pH, alkalinity, apparent colour, concentrations of the ions F^-, Cl^-, S_4^{2-}, Ca^{2+}, Mg^{2+}, Fe^{2+}, Al^{3+}, and hardness. The World Health Organisation recommended limits were employed as bench marks [48]. Table 7 illustrates the problems that were addressed by this research and potential solutions obtained.

Table 7: Major findings in defluoridation with bauxite, gypsum and magnesite at Ngurdoto

Material	Challenge	Result Obtained	Proposed Solution
Bauxite	High turbidity, above 1 NTU, in treated water when used raw	When calcined above 200 °C and employed in fixed bed turbidity reduced to below 1 NTU	May use bauxite calcined at 200 °C in fixed bed to reduce turbidity
	Residual colour beyond 50 TCU	The composite 3:2:1 of bauxite, gypsum and magnesite calcined at 200 °C obtained colours below 50 TCU in fixed bed.	Combine the materials in this ratio and calcine at 200 °C, employ in fixed bed
	Residual Al3+ beyond 0.2 mg/l	The composite 3:2:1 of bauxite, gypsum and magnesite calcined at 200 °C obtained Al3+ Concentrations below 0.2 mg/l in fixed bed.	As above.
Gypsum	Residual hardness above 500 mg/l as CaC_O3	The composite 3:2:1 of bauxite, gypsum and magnesite calcined at 200 °C obtained hardness below 500 mg/l as $CaCO_3$ in fixed bed.	As above.
	High residual $SO_2{-}_4$ beyond 400 mg/l	Gypsum calcined at 400 °C obtained residual sulphates lower than 100 mg/l, composite 3:2:1 calcined at 200 °C employed in fixed bed obtained similar results & higher loading capacity.	May employ calcined gypsum but composite has higher loading capacity therefore composite is better choice to gypsum
Magnesite	Residual pH above 8.5	Composite described above obtained pH between 6.7 and 8.0 in fixed bed	Composite may be used instead of magnesite.

Bauxite obtained from Kwemashai in Lushotho District of Tanzania when characterised for composition showed that the major components were Al_2O_3 (30.33%), SiO_2 (15.00%) and Fe_2O_3 (14.30%). Fluoride removal with bauxite is known to depend mainly on reactions of the Al_2O_3. Oxides of aluminium are amphoteric and will react as base or acid represented in Equations 16 and 17 respectively.

$$Al_2O_3(s) + 6H_3O^+(aq) + 3H_2O(l) \rightarrow 2[Al(OH_2)_6]^{3+}(aq)$$
(16)

$$Al_2O_3(s) + 2OH^-(aq) + 3H_2O(l) \rightarrow 2[Al(OH)_4]^-(aq)$$
(17)

Defluoridation capacity of bauxite decreased with increase in pH of the medium, a result attributed to dominance of acidic reaction of Al2O3 shown in Equation 17 with formation of negatively charged species. The negatively charged species would in effect retard fluoride sorption. Formation of positively charged species as shown in Equation 16 would occur in low pH medium hence greater adsorption of fluoride, fluoride being an anion [46,47].

Contact precipitation of fluoride was also researched on at Ngurdoto [33,34]. The technique employed Calcium chloride and Sodium dihydrogen phosphate and has been described in section 2.4.2. The technology best demonstrated in Tanzania is use of bone char at household level in fixed bed. The configuration designed at Ngurdoto Research Station is shown in Figure 12.

Figure 12: Household defluoridation unit employing bone char developed by Ngurdoto Research Station shown in average and low income household settings.

This column for fixed bed defluoridation has also been scaled up particularly in Arusha national park by the Ngurdoto defluoridation station shown in Figure 13.

Figure 13: Incomplete and completed community defluoridation unit designed by Ngurdoto Defluoridation research station.

Clean raw cow bones are collected and charred at 500 – 550 ° C in specially designed kilns in which air supply is controlled, Figure 14. The charred bones are pulverized to particle sizes of range 0.5 to 3 mm in diameter.

Figure 14: Different sizes of bone charring kilns as developed at the Ngurdoto Defluoridation Research Station.

The bone char has high sorption capacity initially however the media gets saturated with fluoride with time. The general practice around Arusha is to replace the media when the effluent has a fluoride concentration of 2 mg/L, the maximum permissible limit as per International Reference Centre (IRC) in the Hague, Netherland. Results obtained from a typical bone char household unit are shown in Figure 17. The figure shows that defluoridation of water with initial fluoride of 10 mg/L up to about 1300 litres of treated water are obtained with fluoride less than 2 mg/L. The mass of bone char used in this investigation was 4 kg. Based on use of 20 litres per family per day of the treated water, the 4 Kg of bone char can be employed to treat water for sixty five days family use. This approximates to 20 to 25 kg of bone char in a year for a family.

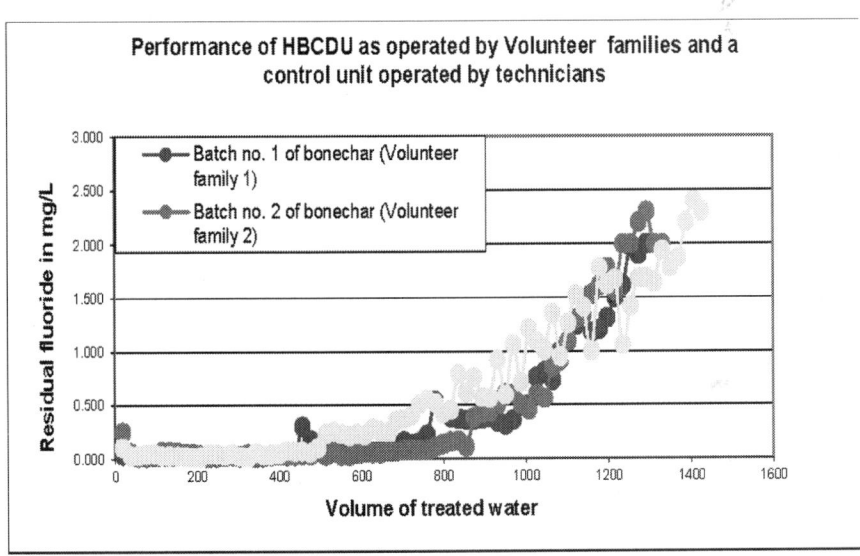

Figure 15: Plot of residual fluoride against volume of treated water for household column units (obtained from Ngurdoto research station development report, 2010).

Investigations have shown that the higher the initial fluoride concentration the higher the bone char exhaustion rate and the smaller amount of bone char the faster the exhausted rate. Fixed bed defluoridation had better performance compared to defluoridation in batch.

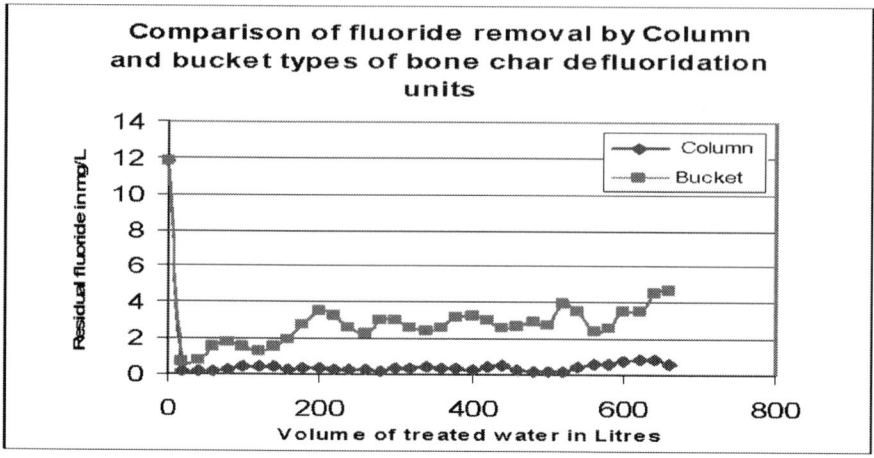

Figure 16: Residual fluoride obtained in column and bucket defluoridation.

Water Defluoridation Research in Malawi

Water defluoridation research in Malawi has been tested with bauxite, gypsum, clay, synthetic and natural hydroxyapatite (HAP) [5, 13, 14]. Gypsum was obtained from Mponela, in Dowa Diastrict of Malawi. Bauxite was obtained from Mulanje Mountain in Mulanje, Malawi and natural hydroxyapatite was obtained from Phalombe District, Malawi. Clay was obtained from Namadzi, Chiradzulu District of Malawi. The materials were calcined at various temperatures, within range 200 to 600 °C, for two hours. Bauxite obtained highest sorption capacity (3.05 mg/g) when calcined at 200 °C. Gypsum obtained highest capacity, 2.17 mg/g, when calcined at 400 °C. Clay obtained highest capacity, 2.15 mg/g, when calcined at 300 °C. Synthetic hydroxyapatite had a fluoride sorption of 1.70 mg/g. Preparation of synthetic hydroxyapatite involved controlled addition of 98 % H_3PO_4 in an aqueous suspension of CaO with periodic additions of 50 % aqueous ammonia. This was followed with decanting the supernatant and then drying the remaining precipitate at 60 °C overnight. The product was then sintered at 1100 °C [49]. The synthetic hydroxyapatite was thus not calcined to various temperatures because its preparation involved high temperature. The natural hydroxyapatite introduced more fluoride in the water with residual fluoride increasing from 8.0 mg/l to 9.65 mg/l. X-Ray

Diffraction (XRD) characterisation showed the material composition were as shown in Table 8.

Table 8: Major compounds in bauxite, gypsum, clay, synthetic and natural HAP tried in Malawi

Raw material	Major compound composition as per JCPDS [50]
Bauxite	$Al_2Si_2O_5(OH)4$
Gypsum	$CaSO_{4.2}H_2O$
Clay	
Synthetic HAP	$Ca_5(PO_4)_3OH$
Natural HAP	$Ca_5(PO_4)_2CO_3(OH)F, Ca_5Al_2(OH)_4Si_3O_{12}, Ca_5(PO_4)_3F$

The natural HAP contained fluoride that explained why it acted as a fluoridating agent. The research also showed that initial quality of the raw water impacts on the defluoridation capacity of the materials tested. Higher initial concentrations of carbonates and chlorides reduced fluoride sorption whereas high initial concentrations of calcium enhanced the sorption in defluoridation with bauxite. In defluoridation with gypsum the higher the initial concentrations of carbonate, nitrate and chloride ions the lower was the sorption of fluoride. Phosphate and chloride interfered with fluoride sorption in defluoridation with synthetic hydroxyapatite.

OVERVIEW OF DEFLUORIDATION TECHNOLOGIES

Table 9 illustrates typical defluoridation outcomes and limitations when different materials are employed. The principals involved in the technologies are briefly outlined in the table to summarise the defluoridation techniques that have been widely demonstrated.

Table 9:

Technology/ Material	Typical capacities (mg/g)	The science	Strengths	Limitations

Activated alumina	3.5–10.0	Precipitations involving Al_2O_3 and in $F-$ ions water	High selectivity for fluoride	lowers pH of water, residual $A^{l3}+$
Nalgonda	0.7 – 3.7	Reactions of Alum, $Al_2(SO_4)3$ and lime (CaO)	Same chemicals used for ordinary water treatment	High chemical dose, high sludge disposal required
Bone char	2.3 – 4.7	Filtration and ion exchange in $Ca_5(PO_4)_3OH$ structure	Availability of raw materials	Not universally acceptable
Bauxite	3.0 – 8.9	Precipitations involving Al_2O_3 and $F-$ and other oxides e.g. Fe_2O_3 ions water	Available locally in some areas. High capacity	Residual colour and turbidity in treated water if used raw
Gypsum	1.1 – 6.8	Ion exchange involving $CaSO_4$ and $F-$ and other compounds e.g. $Ca(OH)2$	Locally available in some areas	High Residual Calcium sulphate
Magnesite	1.0 – 3.7	Ion exchange and precipitation involving MgO and $F-$ and other compounds e.g. $Mg(OH)2$	Simple technique, locally available in some areas	High pH & residual Mg.

HAP	0.5 – 2.9	Ion exchange and precipitation involving $Ca_5(PO_4)_3OH$ and $F-$ and other compounds e.g. $Ca_5H(PO_4)_3(OH)_2$	Naturally available in some areas	Residual Phosphate
Bauxite, gypsum, magnesite composite	4.2– 11.3	ion exchange and precipitation in reactions of $Al_2O_3, CaSO_4, MgCO_3, MgO$	Simple and versatile. Better than use of each of the materials	Energy intensive, fairly novel technique.
Zeolites	28 - 41	ion exchange and surface complexation reactions	High capacity	Limited availability
Other advanced techniques	High	Nano-filtration, Reverse osmosis, distillation, precipitation, electrolysis	Very high capacities	High cost. Need for special training

CHAPTER CONCLUSIONS

There is a wide range of defluoridation techniques and materials to employ when fluoride levels in potable water are likely to result in fluorosis. However the decision on limits of fluoride concentrations in potable water for any region must be guided by average annual daily temperatures, dietary habits, nature and levels of activities in the particular area. Choice of technology will depend on appropriateness where factors such as availability of materials, cost, level of defluoridation required and technical complexity need to be considered. In East and Southern Africa naturally occurring materials such as bone, limestone $(CaCO_3)$, alum (Al_2SO_4), bauxite, gypsum, magnesite, and such other available materials with fluoride affinity, may be given priority when selecting raw materials for treatment and use in water defluoridation. Adoption and/or adaptation of existing technologies require some research at local level because communities differ in socio, economic, religious and traditional status and norms. A basic level of research is therefore paramount to establish appropriateness of a preselected technology to ascertain sustainability of the intervention.

ACKNOWLEDGEMENTS

The author acknowledges the International Programme in Chemical Sciences (IPICS) for supporting his research in Malawi (2002 – 2005), Malawi Government for financing his research in Tanzania (2009 – 2013) The Universities of Malawi and Dar es Salaam, and the Ngurdoto Defluoridation Research Station for guidance, material, technical and moral support rendered.

REFERENCES

1. Ansari M, Kazemipour M, Dehghani M, Kazemipour M. The defluoridation of drinking water using multi-walled carbon nanotubes. Journal of Fluorine Chemistry 2011; DOI: 10.1016/j.jfluchem.2011.05.008.

2. Thole B. Defluoridation kinetics of 200 oC calcined bauxite, gypsum, and magnesite and breakthrough characteristics of their composite filter. Journal of Fluorine Chemistry 2011; 132, 529–535.

3. J. Fawell, K. Bailey, J. Chilton, E. Dahi, L. Fewtrell L, Magara Y., editors. Fluoride in Drinking-water. World Health Organization. London: IWA Publishing; 2006.

4. Sajidu SM, Masumbu FFF, Fabiano E. Ngongondo C. Drinking water quality and identification of fluoritic areas in Machinga, Malawi. Malawi Journal of Science and Technology 2007; 8, 042– 056.

5. Msonda KWM, Masamba WRL, Fabiano E. A study of fluoride ground water occurrence in Nathenje, Lilongwe, Malawi. Physics and Chemistry of the Earth, Parts A/B/C 2007; 32 (15 – 18) 1178 – 1184.

6. Grobler SR, Louw AJ, van Kotze TJ. Dental fluorosis and caries experience in relation to three different drinking water fluoride levels in South Africa. International Journal of Paediatric Dentistry 2001; 11 (5) 372–379.

7. Mjengera H, Mkongo G. Appropriate defluoridation technology for use in fluoritic areas in Tanzania. Physics and Chemistry of the Earth 2003; 28, 1097–1104.

8.   Williamson, MM. Endemic dental fluorosis in Kenya a preliminary report. The East African Medical Journal 1953; 30 (6) 217–233.

9.   Dean HT, Dixon RM, Cohen C. Mottled enamel in Texas. Public Health Reports 1935; 50 (13) 424–442.

10.   United States Public Health Services. PHS Review of Fluoride: Benefits and Risks: Report of Ad Hoc Subcommittee on Fluoride. Committee to Co-ordinate Environmental Health and Related Programs 1991; US Public Health Service.

11.   Brunt R, Vasak L, Griffioen J. Fluoride in groundwater: Probability of occurrence of excessive concentration on global scale. Report SP 2004-2 2004; International Groundwater Resources Assessment Centre (IGRAC).

12.   Nair KR, Manji F. Endemic fluorosis in deciduous dentition – A study of 1276 children in typically high fluoride area (Kiambu) in Kenya. Odonto-Stomatologie Tropicale 1982; 4, 177–184.

13.   Sajidu SMI, Masamba WRL, Thole B, Mwatseteza JF. Ground water fluoride levels in villages of Southern Malawi and removal studies using bauxite. International Journal of Physical Sciences 2008; 3, 001 – 011.

14.   Thole B. Water defluoridation with Malawi bauxite, gypsum and synthetic hydroxyapatite, bone and clay: Effects of pH, temperature, sulphate, chloride, phosphate, nitrate, carbonate, sodium, potassium and calcium ions. MSc thesis. University of Malawi; 2005.

15.   Masamba WRL, Sajidu SM, Thole B, Mwatseteza JF. Water defluoridation using Malawi's locally sourced gypsum. Physics and Chemistry of the Earth 2005; 30, 846–849.

16.   ] Carter GS, Bennet JD. The Geology and Mineral Resources of Malawi. Zomba: Government print; 1973.

17.   Chikte UM, Louw AJ, Stander I. Perceptions of fluorosis in Northern Cape communities. Journal of the South African Dental Association 2001; 56 (11) 528–532.

18.   McCaffrey LP, Willis JP. Distribution and origin of fluoride in rural drinking water supplies in the Western Bushveld Areas of South Africa. In: 4th International Symposium on Environmental Geochemistry 1997, 62; 1997.

19. Mauguhan-Brown H. Our Land. Is our population satisfactory? The results of inspection of school ages. South African Medical Journal 1935; 9, 822.

20. Shrivastava KB, Vani A. Comparative Study of Defluoridation Technologies in India. Asian Journal of Experimental Science 2009; 23 (1) 269-274.

21. Wang Y, Reardon EJ. Activation and regeneration of a soil sorbent for defluoridation of drinking water. Applied Geochemistry 2001; 16, 531–539.

22. Singh R, Maheshwari RC. Defluoridation of drinking water – a review. Indian Journal Environmental Protection 2001; 21 (11) 983–991.

23. Raichur AM, Basu MJ. Adsorption of fluoride onto mixed rare earth oxides. Separation and Purification Technology 2001; 24, 121–127.

24. Bulusu KR, Pathak BN. Discussion on water defluoridation with activated alumina. Journal of Environmental Engineering Division 1980; 106 (2) 466–469.

25. George S, Pandit P, Gupta AB. Residual aluminium in water defluoridated using activated alumina adsorption-modeling and simulation studies. Water Research 2010; 44, (10) 3055-3064.

26. Ong ST, Keng PS, Lee SL, Leong MH, Hung YT. Equilibrium studies for the removal of basic dye by sunflower seed husk (Helianthus annuus). International Journal of the Physical Sciences 2010: 5, (8) 1270-1276.

27. Ko DCK, Lee VCK, John F, Porter JF, McKay G. Improved design and optimization models for the fixed bed adsorption of acid dye and zinc ions from effluents. Journal of Chemical Technology and Biotechnology 2002; 77, 1289–1295.

28. Meenakshi R, Maheshwari C. Fluoride in drinking water and its removal. Journal of Hazardous Materials 2006; B137, 456 – 463.

29. Lee VKC, Porter JF, Mckay G. Modified design model for the adsorption of dye onto peat. Institution of Chemical Engineers Part C - Food and Bioproducts Processing 2001; 79 (C1) 21-26.

30. McKay G, Bino MJ. Adsorption of pollutions onto activated carbon in fixed beds. Water Air Soil Pollution 1990; 51, 33–41.

31. Castillo NAM, Ramos RL, Perez RO., Garcia de la Cruz RF, Aragon-Pin~ a A, Martinez-Rosales JM, Guerrero-Coronado RM, Fuentes-Rubio L. Adsorption of Fluoride from Water Solution on Bone Char. Industrial Engineering Chemistry Research 2007; 46, 9205-9212.

32. Bregnhøj H, Dahi E, Jensen M. Modeling defluoridation of water in bone char columns. In: Proceedings of the First International Workshop on Fluorosis and Defluoridation of Water 18–22 October 1995, Tanzania, The International Society for Fluoride Research, Auckland 1997.

33. Dahi E. Contact precipitation for defluoridation of water. Paper presented at 22nd WEDC Conference, New Delhi, 9–13 September, WEDC 1996.

34. Dahi E. Small community plants for low cost defluoridation of water by contact precipitation. In: Proceedings of the 2nd International Workshop on Fluorosis and Defluoridation of Water. Nazareth, 19–22 November 1997, The International Society for Fluoride Research, Auckland 1998.

35. US EPA. Water Treatment Technology Feasibility Support Document for Chemical contaminants. EPA-815-R-03-004, EPA 2003.

36. Dysart A. Investigation of Defluoridation Options for Rural and Remote Communities. Research Report No 41, The Cooperative Research Centre for Water Quality and Treatment, Salisbury SA 5108, AUSTRALIA 2008.

37. Zakia A, Bernard B, Nabil M, Mohamed T, Stephan N, Azzedine E. Fluoride removal from brackish water by electrodialysis. Desalination 2001; 133, 215 - 233.

38. Giesen A. Fluoride removal at low costs. European Semiconductor 1998; 20 (4) 103-105.

39. Hanemaaijer JH, van Medevoort J, Jansen A, van Sonsbeek E, Hylkema H, Biemans R., Nelemans B, Stikker A. Memstill Membrane Distillation: A near future technology for sea water desalination, Paper presented at the International Desalination Conference, June 2007, Aruba.

40. Aqua-Aero WaterSystems WaterPyramid 2007. http://www.waterpyramid.nl (accessed 21 August 2011).

41. Solar Dew The Solar Dew Collector Sytem 2007. http://www. solardew.com/index2.html (Accessed 26 July 2011).

42. Larsen MJ, Pearce EIF. Defluoridation of Drinking Water by Boiling with Brushite and Calcite. Caries Research 2002; 36, 341-346.

43. Melisa J. Defluoridation of drinking water by adsorption of fishbone. MSc thesis. University of Dar es Salaaam, Tanzania 2001.

44. Bablia K. Studies of water defluoridation using activated carbons and activated carbons loaded separately with Magnesia, Alumina and Calcium. MSc Thesis. Univesrity of Dar es Salaam, Tanzania 1996.

45. Singano JJ. Investigation of the mechanisms of defluoridation of drinking water using locally available magnesite. PhD thesis. University of Dar es Salaam, Tanzania 2000.

46. Thole B, Mtalo FW, Masamba WRL. Effect of particle size on loading capacity and water quality in water defluoridation with 200°C calcined bauxite, gypsum, magnesite and their composite filter. African Journal of Pure and Applied Chemistry 2012; 6 (2) 26-34.

47. Thole B, Mtalo FW, Masamba WRL. Water Defluoridation with 150 – 300 oC Calcined Bauxite-Gypsum-Magnesite Composite (B-G-Mc) filters. Water Resources Management VI, Wit Transactions on Ecology and Environment. 2011; 145, 383 – 393.

48. WHO. Guidelines for Drinking-water Quality, 4th Ed., World Health Organisation: Geneva; 2011.

49. Rahman FF, Bonfield W, Cameron RE, Patel MP. Water uptake of polyethylmethacrylate/tetrahydrofuryl metacrylate polymer systems modified with tricalcium phosphate and hydroxyapatite. Royal London School of Medicine and Dentistry, Queen Mary, University of London: London; 1997.

50. Joint Committee on Powder Diffraction Standards (JCPDS). International Centre for Diffraction Data: Japan; 1997.

Analysis of Quality Mineral Water of Serbia: Region Arandjelovac

Miloš B. Rajković[1], Ivana D. Sredović[1],
Martin B. Račović[1], and Mirjana D. Stojanović[2]

[1]Faculty of Agriculture, University of Belgrade, Belgrade, Republic of Serbia
[2]Institute for Technology of Nuclear and Other Mineral Raw Materials (ITNMS), Belgrade, Republic of Serbia

ABSTRACT

In this paper it is presented the analysis of basic physical and chemical parameters, alkalinity and acidity, the analysis of kations, anions, heavy metals, microbiological analysis and determination of uranium content in waters of Serbia from 10 springs of Arandjelovac region and 2 samples of bottled drinking water. It is done by different methods of analysis according to which conclusion about the content and the quality of these waters can be made. The pH value of analysed waters shows that waters from spring's Maiden spring, Ješovac, Vrelo and Svinčine are slightly acid, while mineral waters from springs Aleksijević, Exploitation and Talpara are slightly basic. The sample from Olga's spring has slightly lower pH value. According to Regulation on the hygiene of drinking water, conductivity should be less than 1000 µS/cm. This condition is fulfilled by waters from springs Aleksijević, Talpara (ordinary), Maiden and Olga's spring, Svinčine and water from city supply system. Springs Ješovac and Vrelo have slightly increased conductivity, while springs Exploitation and Talpara have conductivity significantly above the allowed values (mineral). Analysed natural mineral waters contain only hydrogencarbonates (bicarbonate). According to the content of bicarbonate it was concluded that samples from springs Exploitation, Talpara (mineral), Vrelo and Svinčine belong to the category of bicarbonate waters, as the content of hydrogencarbonate in these samples is higher than 600 mg/dm^3. Analysed mineral waters don't show acidity towards methyl orange, which means that acidity of analysed waters comes from dissolved carbon acid.

INTRODUCTION

Since the beginning of life on Earth till the present days, water has got an immense significance in evolution of life forms, so, it can be said that water is the condition of life preservance on Earth. Vast surfaces covered with water (more than 2/3 of total surface) and total water cubage of about 1400 million km^3, make Earth "the blue planet". But for all its enormous natural wealth, a huge number of population on Earth today is confronted with serious lack of fresh drinking water. From total amount of the world's water reserves, even 97.3% is salty (sea) water (which refining into drinking water is still not economically profitable); 2.7% represents total amount of fresh water, which cannot be fully used. Only 0.3% of all world's water resources is fresh unpolluted water which can be used without any treatment, for drinking or industrial and other purposes [1-3].

Modern society has been characterised by intensive industrialisation and urbanisation with vast exploitation of natural resources and endangering of environment as its consequence. In conditions of global development, care about water is the main issue for civilisation preserving. No matter how enormous are available amounts of water in nature, its usability has been significantly decreased by pollution.

Water is not unlimited natural resources and must be rationally used. So water management and control of water quality become the basic human necessity.

Natural mineral water represents a huge national treasure which can also be used as bottled product, in order to satisfy necessity for drinking water quality. Today we can talk about modern technologic processes of preparation and filling of natural mineral water which represent a whole range of technologic operations which enable a consumer to receive a healthy, natural and quality product. All waters are mineral, table, natural, spring and all are drinkable. They only differ in the degree of mineralisation. The best classification of waters, for consumers, is into table drinking water and table mineral water.

Table drinking waters are all bottled natural spring waters used for drinking and they have no physiological attributes because of the low degree of mineralisation. Table mineral waters have higher degree of mineralisation and they have certain physiologic attributes because of their specific characteristics in content [4].

The purpose of this paper is the analysis of basic physical and chemical parameters, analysis of kations, anions, heavy metals, microbiological analysis and determination of uranium content in waters of Serbia from 10 springs of Arandjelovac region and 2 samples of bottled drinking water. It is done by different methods of analysis according to which the conclusion about the content and the quality of these waters can be made. Waters from springs in Arandjelovac and its surroundings have been used for a long time, in medical purposes, as well as bottled products. Beneficial effects and quality of waters of Arandjelovac region were analysed for the first time by the first chief of medical service in Serbia, Dr Emerich Lindemeier in 1836, and quality of these waters was confirmed by professor Sima Lozanić in 1874.

EXPERIMENTAL

The control of water quality demands representative physical, chemical and microbiologic analysis, as well as adequate choice of quality parameters. To analyse water quality, it is not only significant which parameters will be analysed but also the way of taking samples, its preservation and keeping and starting with the analysis itself. Methodology of taking samples, its frequency, the amount of samples as well as preservation, depends on both quality parameters which will be analysed and the characteristics of water samples. In this paper, a taken water sample was used. At places where the constant flow of spring is present, water samples were taken into 5 dm³ plastic bottles (PET bottles 5L), during the autumn in 2010, and samples were not conserved. A pailful of well water was taken and put into packing material for samples. At springs with faucet, water was left running for 5 minutes before taking a sample. The temperature and conductibility of samples was measured at the place where they were taken.

The analysed water samples, as well as places where the samples were taken are shown in Table 1(and Figure 1). As there is the factory for production of bottled water "Knjaz Miloš" in Arandjelovac, two commercial products of bottled drinking waters were analysed: non-carbonated Aqua Viva and carbonated Knjaz Miloš.

MATERIALS

12 samples of water and 28 parameters of all waters were analysed [5]. The following techniques used for sample analysis: ion chromatography (IC, anions), titration (alkalinity), photometric methods (NH_4^+), potentiometric methods (pH), conductometric method (EC), ion selective electrode (ISE). Detailed analytical procedures are described below. Analytical method and detector limit for all measured parameters are given in Table 2 [6].

Table 1: The analysed water samples and places where the samples were taken

No. of samples	Sample	Places
Sample 1	Waters City	Water supply system of
Sample 2	Aleksijevie well	Arandjelovac
Sample 3	Exploitation	Arandjelovac
Sample 4	Talpara, mineral	Arandjelovac
Sample 5	Talpara, ordinary	Arandjelovac
Sample 6	Maiden spring-water	Arandjelovac Izvor na
Sample 7	Olga's spring-water	Bukulji Izvor na
Sample 8	Ješovac	Bukulji Arandjelovac

Sample 9	Vrelo	Banja village
Sample 10	Svineine	Banja village
Sample 11	Aqua Viva*	Arandjelovac
Sample 12	Knjaz Mild	Arandjelovac

Commercial product—bottled drinking water.

Table 2: Analytical method and detection limit

Parametar	Unit	Analytical method	Detection limit
pH	-	potentiometric	-
EC	pS/cm	conductometric	-
HCO_3^-	mg/L	titration	2
water hardness (total)	M (CaCO3) mg/L	titration	-
As	μg/L	ICP-AES	0.01
Ca^{2+}	mg/L	titration, IC	0.01
Cd	μg/L	ICP-AES	0.001
Cl^-	mg/L	ion-selective electrode, IC	0.01
Cr (total)	μg/L	ICP-AES	0.03
Cu	μg/L	ICP-AES	0.01
F^-	mg/L	ion-selective electrode, IC	0.003
Fe	μg/L	ICP-AES	0.1
K^+	mg/L	IC	0.01
Li	mg/L	IC	0.01
Mg^{2+}	mg/L	titracion, IC	0.01
Mn	μg/L	ICP-AES	0.001
Na^+	mg/L	IC	0.1
Ni	μg/L	ICP-AES	0.01
NH_4^+	mg/L	photometric	0.005

NO_3^-	mg/L	IC	0.01
Sb	µg/L	ICP-AES	0.002
SO_4^{2-}	mg/L	IC	0.01
U	µg/L	fluorimetric	0.0005
Zn	µg/L	ICP-AES	0.05
Alkalinity	mg/L	potentiometric	0.01
Acidity	mg/L	titration	-

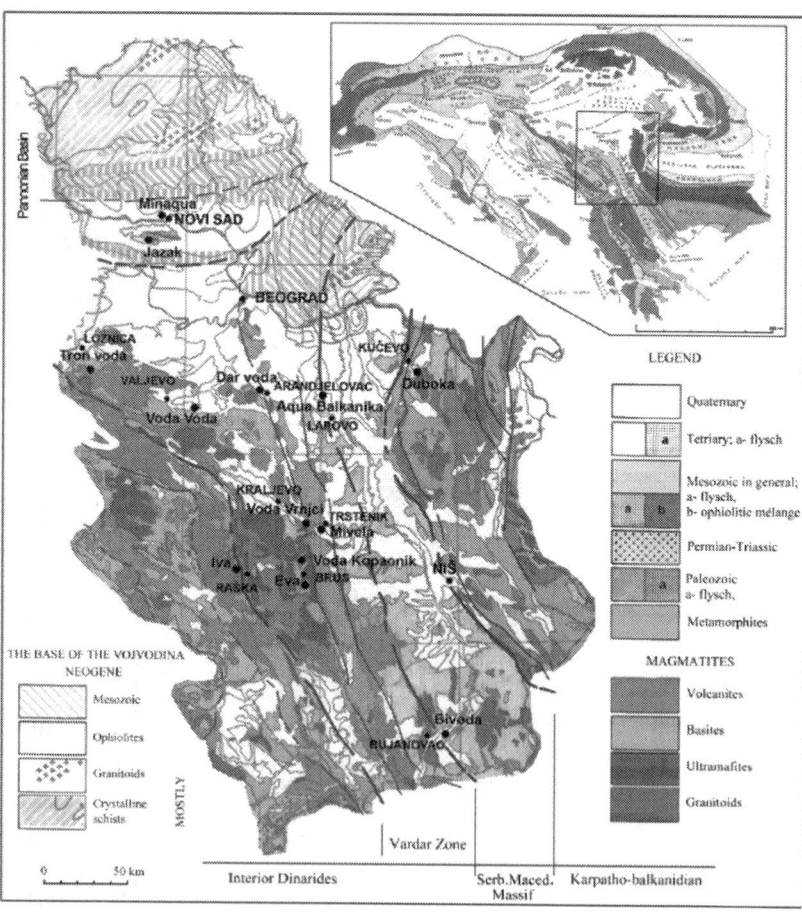

Figure: 1: Geological map of Serbia [13].

Measurements of pH value of solution were done at ion-meter type C863 (Consort, Belgium) [5,7-10].

Conductivity measurements were done at WTW instrument for a terrain multi variant analysis. The instrument was calibrated by standard solutions for calibration (conductivities: 814 µS/cm and 1413 µS/cm) [11].

Determination of alkalinity and acidity of water by volumetric analysis. Water alkalinity was done by conventional volumetric method and potentiometric titration [7]. Before the titration of water samples, solutions of NaOH and HCl were standardised to 0.1 mol/L by primary standard of Na_2CO_3 of 0.05 mol/L concentration [12].

Determination of water hardness. Water hardness was determined in complexometric way, by titration with standard solution of EDTA of 0.01 mol/L concentration.

Determination of dry (solid) residue. Mineralisation (calculated as dry residue at 180°C) represents total mass amount of all present matters in water. First, a glass (mass was previously determined) where samples will be vaporized, was measured at analytical scale. Vaporization of 100 mL of sample is done in water bath till dry. Then the rest in the glass is dried for 8 hours in a drying chamber at 180°C. The glass is cooled and then measured.

Determination of dissolved ammonia [14]. To determine the soluble ammonium UV-VISIBLE SPECTROPHOTO METER SHIMADZU UV-1650 PC was used. The working wavelength was 425 nm.

Solutions received from standard solution of NH_4Cl (T = 0.005 mg/mL) were used for making calibrating plot. The analytic curve is constructed according to measured absorption for said concentration, A = f (c).

The content of ammonia is determined by the equation:

$$x = \frac{c \cdot 50}{V}$$

(1)

where x = content of ammonia (in mg/L); c = ammonia concentration according to the calibration curve (mg/L); V = volume of sample (mL).

Determination of fluoride and chloride ions by ion selective electrode (ISE). Concentration of fluoride and chloride ion in water was measured on ion meter type C863 (Consort, Belgium) and combined fluoride selective electrode (type ISE27B) and combined chloride selective electrode (type ISE24B) was used as a sensor electrodes [15].

For determination of fluoride ion 25.00 mL standard solution of 1000 mg/L concentration (or sample) was measured with pipette into separate flasks, 25.00 mL solution TISAB buffer (Total Ionic Strenght Adjustement Buffer) was added, the magnetic nucleus was inserted and the stirring speed was adjusted.

For determination of chloride ion 25.00 mL standard solution of 1000 mg/L concentration (or sample) was measured with pipette into separate flasks, 25.00 mL solution of KNO_3 of 1 mol/L concentration was added, the magnetic nucleus was inserted and the stirring speed was adjusted.

Ionic chromatography. The instrument used in this analysis is Dionex DX-300. Safety and separation columns for separation of ions are made from same polymeric resin for anion change (IonPac AS14 i AG14-SC), and equipped with anion self regenerating suppressor model ASRS (Auto Self-Regenerating Suppressor) (4 mm) with expected noise conductivity from 5 - 15 mS. Mixture of 2 mmol/L of Na_2CO_3 and 0.5 mmol/L of $NaHCO_3$ was used as mobile phase for separation of anions at this column. Mobile phase flow was 0.8 mol/L, time of the analysis was 15 minutes, and injected cubage was 20 μL.

Polymeric resins for kation exchange (with low capacity of ion exchange) were used for determination of kations. They are equipped with kation self regenerating suppressor model ASRS (Auto Self-Regenerating Suppressor) (4 mm), which separated kations transfer into adequate basis and reduce conductivity of eluent. Methyl sulphonic acid of 20 mmol/L concentration was used as mobile phase for separation of kations. Mobile phase flow was 0.5 mol/L, time of the analysis was 15 minutes, and injected cubage was 10 μL. Samples that had high conductibility, before the analysis by IC method were diluted, and pH value in all analysed waters was adjusted to 3.00 by adding 1 mol/L of HNO_3.

Calibrating plot was made to determine precisely the concentration of anions/kations in the solution. It shows concentrations of standard solutions of anions/kations regarding their corresponding surfaces, and surfaces are calculated at chromatograph made according to their specific conductivities (in mS/cm). Samples and standards were filtrated through membrane syringefilter of 22 µm. Samples and standards were packed into autosampler after filtration.

Inductively coupled plasma atomic emission spectrosopy (ICP-AES). ICP-AES technique is based on the fact that excite atoms emit energy at certain wave-length in their way back into basic condition. Characteristic of each element is to emit energy of certain wave-length which is determined by chemical attributes of the element. Intensity of emitted energy at certain wave-length (characteristic for given element) corresponds to concentration of given element in the sample. As ICP-AES technique is used for determination of substances which are present in traces in a sample, before the analysis, all analysed mineral waters were diluted and acidified by adding the solution of 0.15 mol/L of HNO_3.

Hydride technique. This technique is developed for determination of the elements which make highly vaporised hydrides (As, Hg, Sb, Bi, Se, Ge and Sn) and can be problem when they are directly put into ICP. Knowledge of concentration of these elements in human diet or in natural samples is important for their toxicity. Applying the hydride technique, multiple increasing of sensitivity of determination and separation of elements from complex matrix can be reached.

Gas hydride formed by chemical reaction is transported directly to plasma through gas carrier. The sample, acid and reduction reagents are continually pumped into gas-liquid separator (GLS) from where gas is directly transported into plasma. Levelling entry enables great stability of plasma. Analytic signal is measured continually. Samples are quickly washed and pulled out from GLS. Nowadays, $NaBH_4$ is used as the most common reduction mean, although Sn (II) is used as reduction reagents for Hg.

These elements have to be in appropriate oxidation condition, so in some cases pre-reductions are necessary. As and Sb must be reduced from oxidation condition +5 to +3 to be correctly determined by the hydride technique, while Se must be reduced from oxidation condition +6 to +4. Most common reduction means (for pre-reduction) are ascorbic acid and potassium-iodide.

All measurements were done on instrument Spectro Ciros, according standard method DIN EN ISO 11885 (1998) [16].

Radioactivity. The quantitative content of uranium has been determined by the fluorometric method based on the linear dependence of the fluorescence intensity of uranium solutions on their concentration. The linear dependence occurs within a very large range of low concentrations (to the magnitude of four). The reduction in the fluorescence intensity has been brought to the lowest degree possible by the technique of "standard addition" after the extraction of uranium by the synergetic mixture TOPO (three-n-octal phosphine oxide) ethyl acetate. The fluorescence intensity has been determined by means of the Fluorimeter 26-000 Jarrel Ash Division (Fisher Scientific Company, Waltham, 1978) [17-22].

Micro-biological analysis of water. Total amount of (aerobic mesophilic) bacteria in 1 mL of water was determined in the following way: 1 mL of analysed sample of water and then deep nutritious agar (P1), previously melted and cooled till approximately 50°C, were put by sterile pipette into a Petri dish. Microorganisms in the basis were suspended by easy rotation of Petri dish at flat surface. Planted basis was incubated for 24 hours at 37°C and then colonies were counted.

The result is expressed as cfu/mL (colony forming unit/mL) [9,23].

RESULTS AND DISCUSSION

There are two Regulations in Serbia concerning the quality of water for human use. There are: Regulation on the Hygienic Acceptability of Potable Water (Official gazette of FRY, number 42/98 and 44/99), and Regulation on Quality and Other Requirements for Natural Mineral Water, Spring Water and Bottled Drinking Water (Official Gazette of Serbia and Montenegro, number 53/05).

Comparison of Regulations and Standards in Serbia with EU Directive and World Health organization (WHO, 2006) is shown in the Table 3.

The Regulation on the Hygienic Acceptability of Potable Water defines the maximum acceptable concentrations (MAC) of chemical substances in water for public water supply, and beside that it particularly gives MAC for certain minerals in bottled water (Al, Ba, Ca, Cl^-, CN^-, Cu, F^-, Fe, Hg, Mg, Mn, Na, Ni, NO_2, NO_3^-, SO_4^{2-}, Zn and electro conductivity less than 500 μS/cm). Except for As, B and U, all of the MAC in this Regulations are lower than the one given by the World Health Organization (WHO).

Regulation on Quality and Other Requirements for Natural Mineral Water, Spring Water and Bottled Drinking Water defines the MAC of certain chemical parameters that can be a risk to human health, indicators of water quality and nomenclature of mineral waters. If the concentration than the one given in Table 3 (value **), then this must be highlighted in the water name (on bottle label). Example: "bicarbonate water" if HCO_3^- > 600 mg/L or "Mg water" if Mg > 50 etc.

MAC for F^- (1.5 mg/L), Cl^- (250 mg/L), CN^- (70 µg/L) and SO_4^{2-} (250 mg/L) must not exceeded, while for water rich in CO_2, pH value can be less than 6.80. This Regulation applies to all ground water, regardless of the overall mineralization. The term "spring water" was used meaning "captured water on the location". The Regulation is harmonized with WHO standards, except in cas of B content, for which two times higher maximum concentration is allowed.

The total of 28 parameters were examined for each sample, and some of the results are given inTable 4.

Determination Basic Physical-Chemical Parameters

According to the results shown in Table 4, it can be concluded that water temperature from springs is in expected limits, as they are ground water. Slightly higher temperature is at springs Talpara and Ješovac, because these springs are connected to communal system of exploitation. Water temperature at spring-water Svinčine is also higher than at other spring-waters, because there is accumulation under the spring-water, so this water is in direct contact with the environment. The highest temperature is at spring-water Exploitation, because it is a spring of thermal mineral water.

Value of pH in analysed waters shows that waters from Maiden spring, Ješovac, Vrelo and Svinčine springs are slightly acid, while mineral waters from Aleksijević, Exploitation and Talpara springs are slightly basic. According to Regulation on the hygiene of drinking water (Official Gazette of FRY, number 42/98 and 44/99), only water from Olga's spring-water doesn't satisfy the condition that pH value must be between 6.50 and 8.50.

Bottled water Knjaz Miloš (pH value 5.57) has lower pH value, but it is carbonated water and has increased amount of CO_2, so this low pH value is expected. Received pH values of analysed waters show that lower pH value corresponds to lower content of hydrogen carbonate with the exception of Olga's spring, indicating the presence of other substances with basic attributes.

Concentration of oxygen in water depends on its solubility from air, which depends on the temperature of analysed waters, concentration of dissolved salts, depth of the spring and atmospheric pressure. In analysed samples, concentration of oxygen is expectedly low, as they are ground spring-waters. All samples are according to demands from Regulation on the hygiene of drinking water, as the concentration of dissolved oxygen is far below the maximum allowed value which is 8.24 mg/L. Water at spring Svinčine has the highest content of dissolved

Table 3: Comparison of regulations and standards in serbia with the EU directive and WHO [30]

PARAMETER	UNIT	EU Directive Directive 1998/83EC DRINKING WATER	EU Directive Directive 2009/54/EC NATURAL MINERAL WATER	WHO	Measurement Concentration in Analytical Samples MIN	MAX	Regulation on the Hygiene Acceptability of Potable Water (Official Gazette of FRY number 42/98 and 44/99) MAC of chemical substances in water for public water supply	MAC for certain materials in bottled water	Regulation on Quality and Other Requirements for Natural Mineral Water, Spring Water and Bottled Drinking Water (Official Gazette of Serbia and Montenegro, number 53/05)
pH		6.50 9.50			5.84	7.17	6.80 - 8.50	6.50 - 8.50	
EC at 20°C	µS/cm	2500 g.v.			159	3296	<1000	<500	2500
O_2	mg/l				0.01	0.08			
Alkalinity	M(HCO_3^-) mg/l				69.6	928.3			
Acidity	M($CaCO_3$) mg/l				16.2	516.5			
Water hardness (total)	M($CaCO_3$) mg/l				60.71	964.20			
As	µg/l	10		10	<0.2	3.3	10	50	50
Ca	mg/l	n.d.	>150		36.97 83	155 144	200	300	150
Cd	µg/l	5			0.02	0.94	3	5	1
Cl⁻	mg/l	250 g.v.	>200	300	2.73	54.4	200	25	200 - 250
Cr (total)	µg/l	50		50	1.4	34.5	50	50	50
Cu	µg/l	2000		2000	2.47	73.0	2000	100	2000
F⁻	mg/l	1.5	>1	1.5	0.09 0.14	0.84 0.26	1.2	1	1 - 1.5
Fe	µg/l	200 g.v.		300	2.31	77.90	300	50	200
K⁺	mg/l	n.d.			1.17 1.88	82.99 28	12	10	
Li⁺	mg/l	n.d.			<0.03	5.73			
Mg	mg/l	n.d.	>50		3.89	80.77	50	30	50
Mn	µg/l	50 g.v.			3.31	77.90	50	30	50
Na⁺	mg/l	200 g.v.	>200	200	7.42 11.7	143.74 280	150	20	200
Ni	µg/l	20 g.v.		70	3.8	22.4	20	10	20
NO_3^-	mg/l	50 g.v.		50	<0.05	46.12	50	3	50
Sb	µg/l	5		20	0.18	1.15	3	10	5
SO_4^{2-}	mg/l	250 g.v.	>200	250	0.52	97.64		25	200 - 250
U	µg/l	n.d.		15	<1.0	4.0		50	

Note: n.d.—not detectable; g.v.—groundwater; MAC—Maximum Acceptable Concentrations; *

—titration; **—lower limit of nomenclature; ***—bottled water; a —IC; WHO—World Health Organization.

Table 4: Results of analysed samples

Sample	pH	EC	O_2	Alkalinity*	Acidity	Water hardness	Ca^{**}	Mg^{**}	Fe	Mn	Ni	Cu	Zn
Sample 1	6.69	159	0.03	59.70	16.20	64.30	22.35	3.89	896	24.10	10.70	42.40	248
Sample 2	7.15	830	0.02	351.40	53.80	323.20	76.14	18.52	1256	7.40	8.90	8.70	267
Sample 3	7.22	3853	0.02	5026	516.50	212	23.74	9.47	671	77.30	3.80	10.60	47.10
Sample 4	7.36	5280	0.02	4787	473.40	225	21.13	8.53	664	75.40	4.00	9.60	34.30
Sample 5	7.37	607	0.02	381.30	43.00	296.80	77.45	13.15	1164	3.31	7.10	2.47	34.50
Sample 6	6.30	200	0.01	69.60	59.20	60.71	15.63	5.26	756	5.68	13.20	6.80	110.60
Sample 7	5.84	247	0.03	119.40	107.60	78.57	34.86	5.66	1863	4.94	16.60	10.40	64.50
Sample 8	6.46	1103	0.02	576.80	161.40	473.21	16.97	27.45	1308	8.35	13.40	73.00	37.10
Sample 9	6.33	1174	0.04	928.30	365.50	760.90	45.80	40.77	1646	4.08	15.40	10.60	55.50
Sample 10	6.67	914	0.08	663	107.60	115	101	39.73	1363	4.80	22.40	24.70	41.30
Sample 11	6.92	549	n.d.	364.70	32.30	291	83	19.40	20	n.d.	n.d.	n.d.	n.d.
Sample 12	5.57	1680	n.d.	1512	2690	568.20	144	36.40	70	n.d.	n.d.	n.d.	n.d.

Sample	Cr µg/mL	As µg/mL	Cd µg/mL	Sb µg/mL	Li* µg/mL	Na^+ µg/mL	K^+ µg/mL	F^- mg/mL	Cl^{***} mg/mL	NO_3^- mg/mL	SO_4^{2-} mg/mL	NH_3 mg/mL	Average value of total amount of bacteria in 1 mL	t (°C)	κ^t, µS/cm	κ^{20}, µS/cm	U µg/mL
Sample 1	4.30	3.20	0.30	0.74	<0.05	5.64	1.91	0.11	6.14	0.34	21.24	0.04	-	13	139	159	-
Sample 2	5.00	3.30	0.94	0.70	<0.05	41.54	1.17	0.23	52.53	21.63	57.64	0.33	7	13	728	830	0.004
Sample 3	1.40	<0.20	0.02	0.34	5.67	137.30	82.99	0.84	19.87	<0.05	0.64	0.61	-	31	4940	3853	-
Sample 4	1.40	0.27	0.02	0.18	5.73	143.74	82.14	0.84	20.71	<0.05	0.52	0.74	-	15	4800	5280	-
Sample 5	5.00	2.59	0.23	0.78	<0.05	9.38	1.99	0.14	12.27	7.36	24.17	0.14	225	14	542	607	<0.001
Sample 6	5.00	<0.02	0.33	0.85	0.28	10.96	2.67	0.14	2.73	3.19	27.92	0.08	16	12	172	200	<0.001
Sample 7	4.40	<0.02	0.73	1.10	0.32	15.17	2.56	0.09	5.11	0.43	23.94	0.08	35	12	213	247	0.003
Sample 8	4.90	<0.02	0.40	0.58	0.36	65.15	3.71	0.70	54.4	46.12	52.44	0.28	125	15	1003	1103	-
Sample 9	4.00	<0.02	0.52	0.79	0.31	16.77	3.48	0.30	4.54	3.49	14.65	0.23	-	15	1067	1174	0.002
Sample 10	34.50	0.50	0.59	1.15	<0.05	1.42	14.65	0.11	9.56	12.61	13.67	0.21	4	16	846	914	-
Sample 11	n.d.	n.d.	n.d.	n.d.	n.d.	13.70	1.60	0.14	n.d.	n.d.	n.d.	n.d.	n.d.	-	-	549	n.d.
Sample 12	n.d.	n.d.	n.d.	n.d.	n.d.	286	28	0.26	n.d.	n.d.	n.d.	n.d.	n.d.	-	-	1680	n.d.

Note: n.d.—not detectable; * titration; **IC.

oxygen. This spring-water is placed in rural part of Banja village, with low accumulation, which is in immediate contact with tended land. Thus increased content of oxygen can be the after effect of decay of organic matter [24].

Results of conductivity measurements at the sample taking temperature and at 20°C are shown inTable 4. As thermostat adjustment of the instrument for conductivity measurement wasn't done, temperature of water sample was measured and with approximate correction factor, f, which for water, in temperature range from10 to 25˚C, is 0.02˚C^{-1}, it was calculated to temperature of 20˚C by the equation:

$$\kappa^{20} = \kappa^{t}\left[1 + f\left(20 - t\right)\right]$$

(2)

According to Regulation on the hygiene of drinking water [25] conductivity should be less than 1000 µS/cm. This condition fulfil waters from springs Aleksijević, Talpara (ordinary), Maiden and Olga's spring-water, Svinčine and water supply system. Springs Ješovac and Vrelo have slightly increased conductivity, while springs Exploitation and Talpara (mineral) have values of conductivity significantly above allowed. According to values of conductivity for temperatures of samples taking, water from Exploitation has higher conductivity, which is expected value, as with increasing of temperature rises its conductivity. However, when these values are calculated to the same temperature (at 20˚C), it can be seen that conductivity is 1.4 times higher at spring Talpara (mineral) than conductivity of water from Exploitation spring.

Factory Knjaz Miloš, Arandjelovac, where Aqua Viva and Knjaz Miloš have been bottled is placed in the bottom of Banja village and it is supposed that water which has been processed in the factory and waters from springs Vrelo and Svinčine belong to the same artery. When the values of conductivity and pH values are compared, it can be seen that there are some conductivity deviation and approximately the same pH values.

Determination of Total Content of Mineral (As Dry Residue)

Results of determination of total amount of minerals (calculated as dry residue at 180˚C) are shown in Figure 2.

According to the total mineral amount (in mg/L) waters City, Aleksijević well, Talpara (ordinary), Maiden spring, Olga's spring and Svičine, as well as Aqua Viva bottled water belong to the category of low mineral waters (<500). Waters from springs Ješovac and Vrelo, as well as water Knjaz Miloš belong to natural mineral waters (500

- 1500), while waters from springs Exploitation and Talpara (mineral) belong to natural mineral waters rich with mineral salts (>1500) [26].

Determination of Alkalinity and Acidity

Acidity of natural waters comes from carbonic acid, and sometimes from humic acids. Most of natural waters are alkali, although they can contain free carbonic acid. So, natural waters can contain both acidity and alkalinity, where acidity can only come from carbonic acid.

Hydroxides, carbonates and hydrogen carbonates of alkali and earth alkaline metals, mainly Na, Ca and

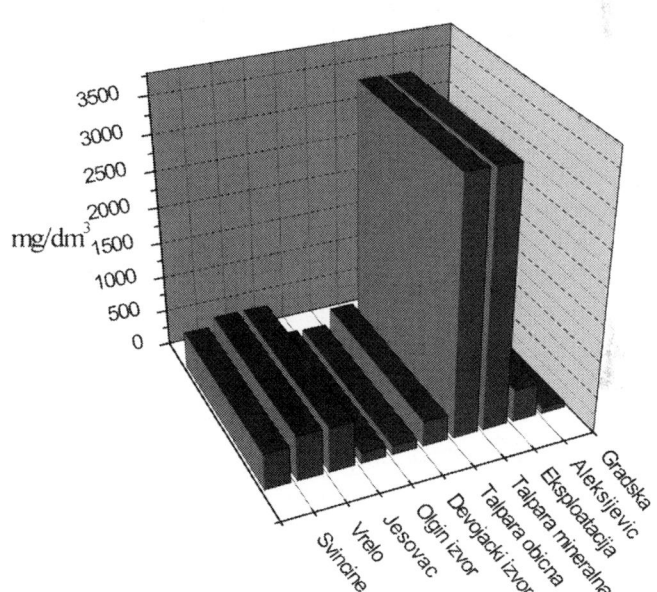

Figure 2: Total content of mineral (as dry residue) in analysed waters.

Mg make water alkalinity. Water alkalinity was determined by the classical volumetric method and the method of potentiometric titration. The results obtained by potentiometric titrations were processed with software Origin 6.1 for math data processing According to the results of alkalinity and acidity determination, it can be seen that all analysed samples contain only hydrogencarbonates, that is, they show only m alkalinity. Results of alkalinity determination by classical volumetric method and by potentiometric titration method are corresponding. Variations are slightly higher with the samples with high alkalinity, as with samples from springs Exploitation and Talpara (mineral).

Acidity towards methyl orange shows the presence of mineral acids, and acidity towards methyl orange at all analysed samples was equal to zero. As there is present only acidity towards phenolphtalein, it can be concluded that acidity of analysed waters comes from dissolved CO_2. The highest acidity has the sample of bottled water Knjaz Miloš, where the content of CO_2 is higher, as it is carbonated water.

Determination of Water Hardness

The results of determination of total and carbonate water hardness show that samples from City supply system, Aleksijević well and Maiden spring has total hardness of water higher than carbonate one. That means that these waters contain only hydrogencarbonate of calcium and magnesium. Small differences in total and carbonate hardness of these samples show that sulphates and chlorides of calcium and magnesium in these samples are present in small concentration. The smallest hardness has the sample from spring at mountain Bukulja, which can be the consequence of low layer of underground accumulation of these waters. In its way towards the spring, water from these springs has relatively small contact with minerals of calcium and magnesium, and at the same time low amount of CO_2, so there is no reaction of making hydrogencarbonates.

As the result of reaction of hydrogencarbonate making it comes to the increasing of amount of Ca^{2+} and HCO_3^- ions in water, that corresponds to alkalinity determination, as alkalinity in these natural waters is the smallest. Here the City supply system should be excluded because it is previously treated and is used for water supply of Arandjelovac.

Samples of other analysed mineral water have carbonate hardness higher than total hardness, and same trend is present with bottled waters. That means that in water besides carbonates and hydrogencarbonate of earth alkaline metals there are higher and bigger amounts of carbonates and hydrogencarbonates of alkaline metals.

The highest hardness, meaning the highest amount of dissolved salts of calcium and magnesium, as well as hydrogencarbonates, have samples Exploitation and Talpara (mineral) and these results correspond to conductivity determination (Table 4).

According to the received results of water hardness analysis (in mg $CaCO_3/L$) it can be concluded that City supply system water and water from Maiden spring and Olga's spring belong to the category of soft waters (<200). Water from Aleksijević well, Talpara (ordinary) as well as bottled water Aqua Viva belong to the category of medium hard waters (200 - 400). Water from spring Ješovac belong to the category of hard waters (400 - 600), while waters from springs Exploitation, Talpara (mineral), Ješovac, Vrelo as well as bottled water Knjaz Miloš belong to the category of very hard waters (>600).

Determination of Cations

Results of determination of Ca and Mg by the method of ionic chromatografy (Figure 3); heavy metals by ICPAES method; uranium by the fluorimetric method and Li^+, K^+ and Mg^{2+} ions by ionic chromatography are shown in Table 4.

Comparative analysis of the results of the analysis of calcium and magnesium amount by the classical titrimetric method and the method of ionic chromatografy, shows the significant deviation from results received from the usage of the methods mentioned above. The reason is high content of iron-ions during volumetric determination of Ca^{2+} and Mg^{2+} ions which, together with Cd^{2+}, Cu^{2+}, Zn^{2+} and Pb^{2+} ions, interfere into complexometric titration of Ca and Mg. Disturbances coming from these ions should be removed before the titration of samples by adding solution of hydroxylamine-hydrochloride. The smallest deviation was noticed during determination of Ca and Mg in mineral water from Maiden spring, where there was the smallest amount of Fe, as the argument to this clam.

The ratio of Ca and Mg in organism is approximately 3:1, so their entry through water should be approximately 3:1. Said ratio is present in mineral water from Maiden spring (2.97) while in other waters it goes from 0.62 to 6.15. Named concentration of elements can be changed indicating that mineral water has no constant composition.

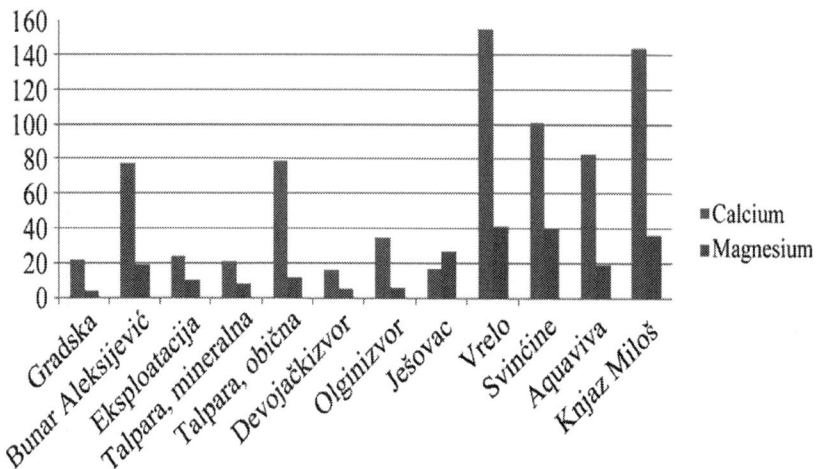

Figure 3: Graphical display of ratio of Ca and Mg amounts in analysed mineral water.

Amount of Fe in analysed waters goes from 664 to 1863 μg/L, which is above MAC values (300 μg/L). Although Maiden and Olga's springs are placed at mountain Bukulja, the amount of Fe in water from Olga's spring is double higher (1863 and 756 μg/L, respectively), indicating that mineral waters from these springs have contacts with different mineral matters [25].

Concentration of Fe and Mn in mineral waters from Exploitation and Talpare (mineral) are almost equal (671 and 77.30 i.e. 664 and 75.4 μg/L, respectively). When other parameters are compared (conductivity, water hardness, dry residue) it can be seen that mineral waters from these two springs are of approximately same content. The analysis of results of heavy metals determination (Table 4) shows that the content of analysed metals is below MAC values. In mineral water from spring Svinčine, the amount of Cr (34.5 μg/L) is slightly higher, while Ni is present in concentration (22.4 μg/L) which is above the maximum allowed concentration (20 μg/L).

The results of ion determination of alkali metals in analysed mineral waters (Table 4) show that the highest concentration of Na^+ ions is present in mineral waters from Exploitation and Talpare (mineral), but it is below the maximum allowed amount. The amount of K is also the highest in samples from Exploitation and Talpare (mineral) and is 6.5 times higher than the maximum allowed concentration (82.99 and 82.14, respectively in comparison to 12 mg/L that is MAC value). Slightly above MAC is the content of K in sample from spring Svinčine (14.65 mg/L), while in other analysed waters the content of K, and also Na is below MAC values (Figure 4).

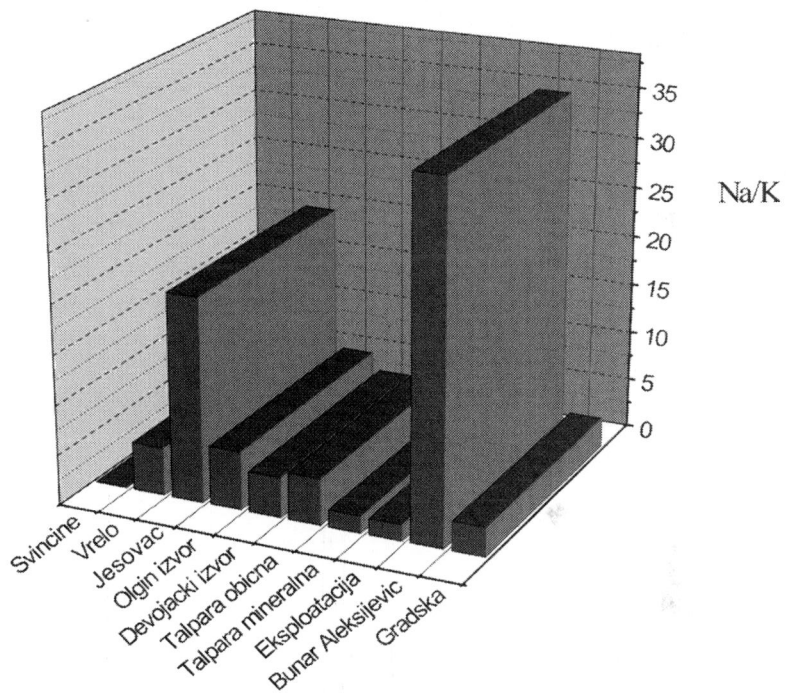

Figure 4: Graphical display of ratio of Na and K amounts in analysed mineral water.

For regular functioning of sodium-potassium pump in cells of human organism it is necessary that ratio of Na and K is 3:2, that is, 1.5 times higher amount of Na than K. If it is supposed that daily entry of these elements through human diet is constant and equal to 1.5, then it would be optimal that entry through water be the same in order not to disturb the balance in organism. For normal functioning of organism it is desirable to drink water with less than 150 and in some cases less than 50 mg/L of Na and with less than 12 mg/L of K.

Determination of Fluoride and Chloride Ions

The measured values of potential in analysed samples of mineral waters with appropriate ion-selective electrode are located in curved part of calibration plot. This is the reason why amount of fluoride and chloride ion is determined by the method of standard addition [8,27].

First, the potential of analysed sample of known cubage was measured (V_t). If Nernst equation is used it can be written that:

$$E_t = k_e + k \cdot \log \gamma_t c_t$$

(3)

where k stands for line inclination which represents dependence of E from log c, γ_t and c_t stand for coefficient of activity, that is, concentration of fluoride-ion in analysed solution. Known cubage V_2 of standard solution of concentration c_s, of F^- or Cl^- ion, was added into the analysed solution and potential E_2 was measured. Concentration of c_s was chosen to be 50 - 100 times higher than concentration of c_t. For potential E_2 is:

$$E_2 = k_e + k \cdot \log \frac{\gamma_t \left(V_t c_t + V_2 c_s \right)}{\left(V_t + V_2 \right)}$$

(4)

where V_t is the volume of analyzed solution.

Making the similar ion volume of both solutions, coefficient of activity in both solutions will be the same, so the difference of the potentials can be expressed as:

$$\Delta E = E_2 - E_t = k \cdot \log \frac{\left(V_t c_t + V_2 c_s \right)}{c_t \left(V_t + V_2 \right)}$$

(5)

It follows that

$$C_t = \frac{C_s}{10^{\Delta E/k} \cdot \left(1 + \dfrac{V_t}{V_2} \right) - \dfrac{V_t}{V_2}}$$

(6)

So, according to inclination of calibration plot known, concentration of F- and Cl- ions were calculated [28], and results are shown in Table 4, where the results of F- and Cl- ions determination by the method of ionic chromatography are also shown.

The analysis of fluoride amount by the method of ionic chromatography in all samples was below the level of detection (0.05 mg/L). The results of measurements by ion-selective electrode show that amount of fluoride is below MAC values in all analysed mineral waters. Namely, the level of detection of fluoride-selective electrode is lower than by the method of IC and is 0.01 mg/L, indicating that ISE method is more sensitive to fluoride determination.

On the other hand, high concentration of Fe means that in analysed samples can come to building of dissolved complexes of iron and F- ions. These complex compounds are less stable from complexes which iron builds with citrate-ion from TISAB solution. By adding the TISAB solution, F-ion frees from the complex which it makes with iron ion and goes into solution. That enables more sensitive and more precise measurements of fluoride ion-selective electrode, in comparison to IC method.

From the analysis of results shown in Table 4, it can be seen that the content of fluoride and chloride too, nitrate and sulphate is nearly same in samples Exploitation and Talpara (mineral).

The analysis of chloride by IC method and ion-selective electrode shows considerate differences in results received by these methods. Approximate values of chloride amount have been found only in samples from Aleksijević well and from spring Ješovac. Obstructions during the work with chloride-selective electrode can come from interfering effect of I-, Br-, OH-, CN-, $S_2O_3^{2-}$, NO_3^- and SO_4^{2-} ions [7,29]. On the other hand, level of detection of chloride determination by ion-selective electrode is 2 mg/L, and by IC method is 0.01 mg/L.

The amount of nitrite, phosphate and bromide ion was also analysed in mineral waters. Concentration of these ions in all samples was below the level of detection.

Determination of Ammonia Content and Micro-Biological Analysis

The analysed results of determination of ammonia amounts and the microbiological analysis show that the amount of ammonia is above MAC values in samples Exploitation and Talpara (mineral). Comparing pH values and content of ammonia, it can be seen that in natural mineral waters lower value of pH (a more acid sample) correspond to lower level of ammonia. So the lowest amount of ammonia is in sample from Maiden spring, which also has the lowest pH value. Here we didn't consider City supply system water because it is chemically and microbiologically treated, and according to Regulation on the hygiene of drinking water.

Drinking water from public springs of closed type can be classified into several categories [25]: refined water (it cannot have coliform bacteria in 100 mL, and total amount of living bacteria is ≤10 in 1 mL of water), natural water (the content of coliform bacteria ≤10 in 100 mL, and total amount of living bacteria is ≤100 in 1 mL of water).

Drinking water from opened wells and other public springs which are used for drinking can contain up to 300 of total bacteria in 1 mL. As for total number of bacteria in mineral waters from springs Talpara (ordinary) and Ješovac, they have slightly higher amount of bacteria comparing to other analysed waters. However, these two springs belong to the category of public springs, meaning that microbiologic quality of waters from these springs is according to the Regulation.

CONCLUSIONS

The pH value in analysed waters shows that waters from springs Maiden spring, Ješovac, Vrelo and Svinčine are slightly acid, while mineral waters from springs Aleksijević, Exploitation and Talpara are slightly basic.

In analysed waters concentration of oxygen is expectedly low but all samples are according to Regulation on the hygiene of drinking water (MAC values 8.24 mg/L).

Conductivity is satisfied with waters from springs Aleksijević, Talpara (ordinary), Maiden and Olga's spring, Svinčine and water from supply system. Springs Ješovac and Vrelo have slightly increased conductivity, and Exploitation and Talpara (mineral) have significantly higher conductivity values than MAC.

Waters City, Aleksijević well, Talpara (ordinary), Maiden spring, Olga's spring and Svičine, as well as bottled water Aqua Viva belong to the category of natural low mineral waters. Samples from springs Ješovac and Vrelo, as well as water Knjaz Miloš belong to the category of natural mineral waters. Samples from springs Exploitation and Talpara belong to natural mineral waters rich with mineral salts (mineral).

Determination of water hardness shows that City water and water from Maiden and Olga's springs belong to the category of soft waters; water from Aleksijević well, Talpara (ordinary) as well as bottled water Aqua Viva belong to the category of medium hard waters; water from springs Ješovac belong to the category of hard waters, while waters from springs Exploitation, Talpara (mineral), Ješovac, Vrelo as well as bottled water Knjaz Miloš belong to the category of very hard waters.

The analysis of Ca and Mg amount shows that analysed mineral waters contain Ca and Mg in concentrations below MAC values.

The amount of Fe in analysed mineral waters is increased and higher than MAC values.

The analysis of the results of heavy metals determination shows that the amount of analysed metals is below MAC values. In mineral water from spring Svinčine, there is slightly increased amount of Cr, while Ni is present in concentration above MAC values.

The highest concentration of Na^+ ions is present in mineral waters from Exploitation and Talpare (mineral), but it is below MAC values. Amount of K is also the highest in samples from Exploitation and Talpare (mineral) and they are 6.5 times higher than MAC values. Slightly above MAC is the amount of K in sample from spring Svinčine, while in other analysed waters amount of K and Na is below MAC values.

Amount of fluoride, chloride, nitrate and sulphate in analysed mineral waters is below MAC values.

Amount of ammonia above MDK values was in samples Exploitation and Talpara (mineral). Comparing pH values and ammonia amount, it can be seen that in natural mineral waters lower pH value (a more acid sample) correspond to lower ammonia amount. The lowest ammonia amount is in the sample from Maiden spring, and it also has the lowest pH value.

The results of analysis show that concentration of uranium is in natural values range.

ACKNOWLEDGEMENTS

We gratefully acknowledge the financial support from the Ministry of Science, Technology and Development, the Republic of Serbia (grant number III 43009) for the research work.

REFERENCES

1. D. K. Todd, "The Water Encyclopedia," Water Information Center, New York, 1970.

2. L. Gavrilović and M. Lješević, "Voda kao Uslov Života i Prirodni Resurs," Zbornik radova sa Konferencije VODA ZA 21 VEK, Udruženje za tehnologiju vode i sanitarno inženjerstvo, Beograd, 1999 (in Serbian).

3. http:// www.mineralwater.org

4. B. Dalmacija, "Kontrola Kvaliteta voda u Okviru Upravljanja Kvalitetom," Prirodno-Matematički Fakultet, Novi Sad, 2000 (in Serbian).

5. Anonym, "Vogel's Textbook of Quantitative Chemical Analysis," 5th Edition on CD-ROM, John Wiley & Sons, Inc., New York, 1989.

6. Environmental Protection Agency (EPA), "Analytical Methods Approved for Drinking Water Compliance Monitoring og Inorganic Constituents National Primary Drinking Water Regulations," CFR 141.23 and Appendix A to Subpart C of Part 141, USA, 2009.

7. M. B. Rajković and I. D. Sredović, "Praktikum iz Analitičke Hemije," Poljoprivredni Fakultet, Zemun, 2009 (in Serbian).

8. D. C. Harris, "Quantitative Chemical Analysis," 6th Edition, W. H. Freeman and Company, New York, 2003.

9.   A. D. Eaton, L. S. Clesceri and A. E. Greenberg, "Standard Methods for Examination of Water and Waste Water," 19th Edition, American Public Healt Association, Washington, 1995.

10. Anonym, "Standard Methods 4500-B+B: Standard Methods for the Examination of Water and Wastewater," 21st Edition, 2005. http://www.standardmethods.org

11. Anonym, "Standard Methods 2510 B: Standard Methods for the Examination of Water and Wastewater," 21st Edition, 2005. http://www.standardmethods.org.

12. Anonym, "Standard Methods 2320 B: Standard Methods for the Examination of Water and Wastewater," 21st Edition, 2005. http://www.standardmethods.org.

13. M. Dimitrijević, "Geological Map 1:2000000. Geological Atlas of Serbia, No. 1," 1994 (in Serbian).

14. Anonym, "Određivanje Sadržaja Amonijaka. Metoda Pomoću Nessler-Ovog Reagensa," SRPS 7150-1, 1990 (in Serbian).

15. Anonym, "Standard Methods 4500-F-E: Standard Methods for the Examination of Water and Wastewater," 21st Edition, 2005. http://www.standardmethods.org

16. M. Birke, C. Reimann, A. Demetriades, U. Rauch, H. Lorenz, B. Harazim and W. Glatte, "Determination of Major and Trace Elements in European Bottled Mineral Water—Analytical Methods," Journal of Geochemical Exploration, Vol. 107, 2010, pp. 217-226.doi:10.1016/j.gexplo.2010.05.005

17. M. Stojanović and Z. Martinović, "Pregled Analitičkih Metoda za Određivanje Urana. Uticaj Upotrebe Fosfornih Đubriva na Kontaminaciju Uranom," Zbornik radova sa naučnog skupa, SANU, Beograd, Knjiga 5, 1993, pp. 19-29 (in Serbian).

18. M. Stojanović and M. B. Rajković, "Određivanje i Karakterizacija Urana u vodi za piće," XXII Simpozijum Jugoslovenskog Društva za Zaštitu od Zračenja, Petrovac n/m, 29.09.-1.10.2003.god., Sekcija 4: Radioekologija, Zbornik radova, pp. 153-156 (in Serbian).

19. M. B. Rajković, M. Stojanović, Č. Lačnjevac, D. Toš- ković and D. Stanojević, "Određivanje Tragova Radioaktivnih Supstanci u Vodi za Piće," Zaštita materijala, Vol. 49, No. 4, 2008, pp. 44-54 (in Serbian).

20.   Službeni list SRJ br. 9/99, "Pravilnik o Granicama Radioaktivne Kontaminacije Životne Sredine i o Načinu Sprovođenja Dekontaminacije," 1999 (in Serbian).

21.   M. B. Rajkovic, C. Lacnjevac, N. R. Ralevic, M. D. Stojanovic, D. V. Toskovic, G. K. Pantelic, N. M. Ristic and S. Jovanic, "Identification of Metals (Heavy and Radioactive) in Drinking Water by an Indirect Analysis Method Based on Scale Test," Sensors, Vol. 8, No. 4, 2008, pp. 2188-2207. doi:10.3390/s8042188

22.   Anonym, "Određivanje Sadržaja Urana Fluorimetrijskom Metodom DM 10-0/34," Institut za Tehnologiju Nuklearnih i Drugih Mineralnih Sirovina (ITNMS), Beograd, 2004 (in Serbian).

23.   M. Vrvić and G. Gojgić-Cvijović, "Praktikum za Mikrobiološku Hemiju," IHTM, Centar za Hemiju, Beograd, 2003 (in Serbian).

24.   I. Panić, "Ispitivanje Opštih Fizičkohemijskih i Radiohemijskih Osobina Voda Mataruške, Selters i Vrnjačke Banje," Diplomski Rad, Fakultet za fizičku hemiju, Beograd, 2009 (in Serbian).

25.   Službeni list SRJ br.42/98, "Pravilnik o Higijenskoj Ispravnosti Vode za piće," 1998 (in Serbian).

26.   Anonym, "Pravilnik o Kvalitetu Prirodne Mineralne Vode," Sl. list SRJ, br. 45/93 i 76/93, ispr. i Sl. list SCG, br. 56/2003, dr. pravilnik, 4/2004, dr. Pravilnik i 37/2005 (in Serbian).

27.   M. B. Rajković and I. D. Novaković, "Priručnik za Upotrebu Fluorid-Selektivne Elektrode u Analizi Biološ- kog Materijala," Poljoprivredni Fakultet, Beograd, 2007 (in Serbian).

28.   http://www. csrg.ch.pw.edu.pl

29.   M. B. Rajković and B. Vučurović, "Selectivity of Copper (I)—Sulphide in the Presence of Different Interfering Ions," Review of Research Work at the Faculty of Agriculture, Vol. 37, No. 2, 1992, pp. 147-154.

30.   T. Petrović, M. Zlokolica-Mandić, N. Veljković and D. Vidojević, "Hydrogeological Conditions for the Forming and Quality of Mineral Waters in Serbia," Journal of Geochemical Exploration, Vol. 107, 2010, pp. 373-381.

Conceptual Hydrogeological Model and Groundwater Resource Estimation in a Complex Hydrothermal Area: The Case of the Viterbo Geothermal Area (Central Italy)

Antonella Baiocchi, Francesca Lotti,
and Vincenzo Piscopo

Dipartimento di Scienze Ecologiche e Biologiche (DEB), Università degli Studi della Tuscia, Viterbo, Italy

ABSTRACT

The conceptual hydrogeological model of the Viterbo thermal area in central Italy and the yield of the groundwater system have been examined. This area is of great geothermal interest. Through new investigations, three overlapping aquifers have been found. This study examines in detail the two shallower aquifers, characterized by different hydraulic and chemical characteristics. The first aquifer is related to the regional groundwater flow of the Cimino-Vico volcanic system and is generally characterized by cold, fresh waters used for irrigation and drinking water supply. The second aquifer, i.e. the thermal aquifer, supply thermal spas and public pools; it is present where the local hydrostratigraphic, structural and geothermal conditions permit a relatively active flow of higher salinity thermal waters (40°C - 62°C). These two aquifers interact vertically and laterally, giving rise to mixed waters circulating in the first aquifer. The first aquifer is recharged by direct infiltration and inflow from regional groundwater, as well as inflow from the second aquifer. The yield of the thermal aquifer is at least 170 L/s, discharging into thermal springs and wells, besides feeding the shallow aquifer vertically and laterally. Even if a future development of the second aquifer is potentially achievable on a global scale, the exploitation of the thermal waters is strictly dependent on the specific local hydrogeological equilibrium between the overlapping aquifers, different from place to place. The case study highlights that, in the volcanic hydrogeological environment, one of the most stringent constraints in determining the correct usage of a resource is the variable level of interaction of groundwater with different qualities.

INTRODUCTION

The city of Viterbo is in the Tuscany-Latium geothermal region and is of great geothermal interest [1]. Several thermal springs and wells are present that have water temperatures as high as 62°C. Some of these springs have been known for their therapeutic properties since Roman times, and wells have been drilled for geothermal exploration since the 1950s.

Currently, these thermal waters are used primarily to supply thermal spas and public pools. In the same area, a shallow aquifer carries cold and fresh water, which is used for irrigation and the local drinking water supply. Increases in spa tourism and the use of geothermal energy are expected in the near future. This multi-purpose water demand also exists for other volcanic aquifers in Italy and around the world e.g., [2-7]. To address the future groundwater management in these complex systems, it is important to examine the local response of the aquifers to withdrawals.

The purpose of this paper is to review the conceptual hydrogeological model of the Viterbo thermal area and to estimate the yield of the groundwater system. This study represents a first step toward determining criteria for the sustainable management of groundwater in this area using a numerical model. Various studies of the Viterbo thermal area have addressed the reconstruction of the stratigraphy and structure, the evaluation of heat flow, the chemical characteristics of gaseous and hydrothermal emissions, and the origin of the thermal waters. However, because of the lack of complete data, few hydrogeological studies have defined the interactions between aquifers and the yield of the system. These aspects are studied herein through new investigations using an integrated approach that combines a hydrogeological and hydrochemical characterization of the complex groundwater system.

GEOLOGICAL AND HYDROGEOLOGICAL OUTLINES

The study area lies between the Tyrrhenian coast and the Apennine Mountains (Figure 1(a)). This region contains a series of sedimentary basins related to the processes that occurred during the formation of the Apennine Chain. Periods of local subsidence alternated with periods of differential uplift and intense volcanic activity have affected the region since the Pliocene [8-11]. Volcanic activity gave rise to the Cimino and Vico complexes (Figure 1(a)); the former is related to the Tuscan-Roman anatectic magmatic province, and the latter is related to the Roman-Campanian potassic alkaline province [12-15].

The Cimino complex was active between 1.35 and 0.95 Ma. Effusive and explosive activity gave rise to several domes that developed along a NW-SE trending fracture and included pyroclastic deposits. Rhyodacitic ignimbrites and domes as well as latitic and olivinelatitic lavas constitute the volcanic complex [16-18].

The Vico complex consists of a stratovolcano with a central caldera that houses Vico Lake. This volcano was mainly active between 419 ka and 95 ka and developed along a NW-SE elongated graben at the intersection with a NE-SW fracture. Alternating explosive and effusive phases gave rise to several pyroclastic deposits and lava flows, which are phonolitic, tephritic and trachytic in composition [18-21].

The thickened folded and thrusted substratum beneath the Cimino and Vico volcanics is composed of MesozoicCenozoic carbonate sequences (that are several thousands of meters thick) and siliciclastic turbidite deposits (the Upper Cretaceous-Eocene flysch) [9,10,22-25]. NWand NE-striking extensional faults subdivide the substratum rocks and control the horst and graben pattern. Neogene-Quaternary marine to continental deposits fill the structural low of the Mesozoic-Cenozoic units.

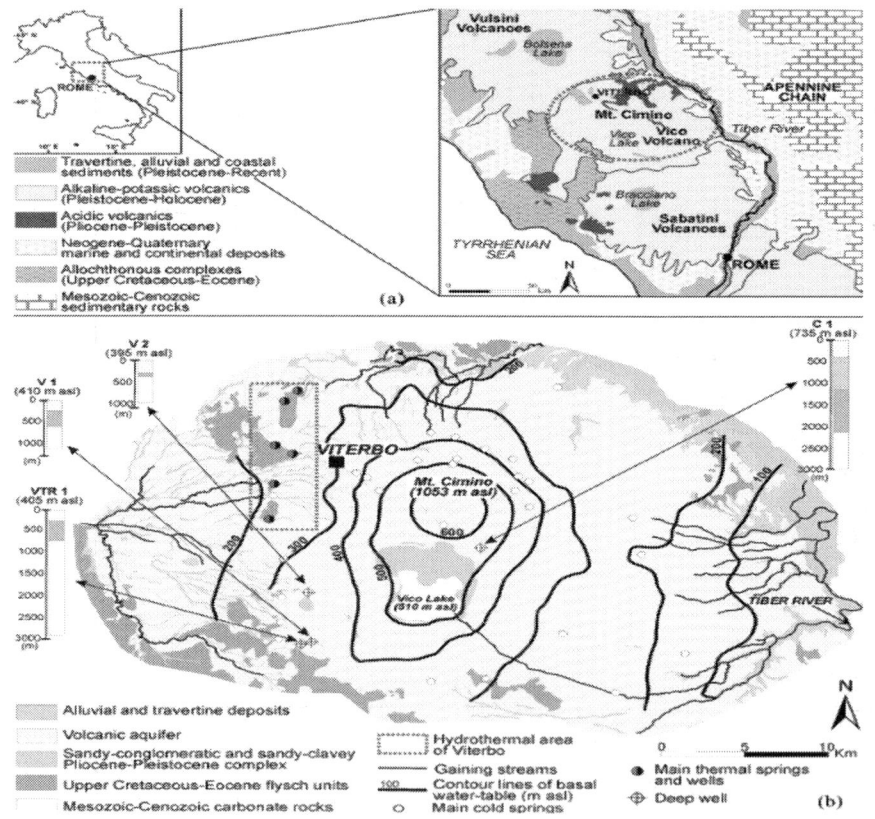

Figure 1: (a) Cimino-Vico volcanoes location in central Italy; (b) Simplified hydrogeological map of the Cimino-Vico system [34] with location of hydrothermal area of Viterbo.

The thinning of the lithosphere and the related igneous processes affecting the pre-Apennine belt explain the formation of the strong regional heat flow anomaly. Values of heat flow between 200 and 300 mW/m^2 over wide areas and up to 400 mW/m^2 in smaller zones have been recognized [1,26-28].

Substantial CO_2 emissions characterize the area and control the genesis of the travertine [29-33] that typically outcrops around the Viterbo thermal area (Figure 1(b)).

The Cimino and Vico volcanites constitute an aquifer system limited by the Pliocene-Pleistocene sedimentary complex and Upper Cretaceous-Eocene flysch units (Figure 1(b)). A continuous basal aquifer and several limited, discontinuous perched aquifers are present [34-36]. The volcanic aquifer discharges mainly into streams and springs, and it flows towards the alluvial aquifer. The mean yield of the volcanic aquifer has been estimated to be between 5 and 7 m^3/s [34].

In the study area, the Mesozoic-Cenozoic carbonate rocks are considered to be a deep aquifer hosting a thermal reservoir [26,37-38]. The shallow and deep aquifers are separated by thick low-permeability Pliocene-Pleistocene and Upper Cretaceous-Eocene sedimentary rocks. West of Viterbo, the uplift of the basement of the volcanites and the high heat flow are considered to be the origin of the uprising of thermal waters via faults and fractures (Figure 1(b)). Sulfate-alkaline-earth-type thermal waters with temperatures between 50°C and 62°C are more mineralized and have higher gas contents (CO_2 and H_2S). By contrast, the waters of the volcanic aquifer comprise fresh and cold bicarbonate-alkaline-earth waters [37,39, 40].

METHODS AND DATA

Based on the present knowledge of the Viterbo thermal area, new investigations were planned, including: 1) hydrostratigraphic data acquisition; 2) flow and water level measurements; 3) pumping tests; 4) chemical and isotopic analyses; and 5) meteorological, soil and land use data processing.

The hydrostratigraphy of the area was reconstructed based on available studies and the interpretation of 62 lithologic logs. These logs concern wells and boreholes with depths ranging from tens to hundreds of meters and report information on the stratigraphy, aquifer formations, water level and, in some cases, water temperature.

Flow measurements were conducted in August-October 2008, May-June 2009 and September 2010 for 15 thermal springs, 9 cold springs, 7 flowing thermal wells and 28 stream sections. Measurements with an accuracy of 5% to 10% were obtained using tanks or current meters.

The water level, temperature and electrical conductivity in the wells were measured in August-October 2008 and May-June 2009 with a multiparametric probe. For the flowing wells, the water level was determined according to the fluid pressure measured with a manometer. In total, 130 wells with depths of 5 to 150 m were measured in or near the hydrothermal area.

Pumping test results at five wells that penetrate the shallow volcanic aquifer were acquired from the literature. Two new pumping tests were performed to monitor the temperature and/or electrical conductivity of the pumped water. Six other step-drawdown tests were also conducted for the same aquifer.

Three new pumping tests were also performed on the thermal wells. Two tests were conducted at a constant rate with observation piezometers or springs to monitor the chemical and physical characteristics of the water at each point. A third test was conducted by opening and closing a flowing well and observing the response of a second well.

Water from a total of 52 sources was sampled during a survey conducted in June 2009 and during the pumping tests.

The temperature, pH and electrical conductivity were measured in the field using portable meters. The alkalinity was determined on-site by means of titration.

Major anions (Cl^-, SO_4^{-2}), nitrate (NO_3^-) and fluoride (F^-) were determined by ion chromatography using a Dionex-DX-120 system. Major cations (Na^+, K^+, Ca^{2+}, Mg^{2+}), Sr, Li and Fe liquid were determined by atomic absorption spectrophotometry with a Perkin-Elmer 2100 system. SiO_2 was determined with a Secomann S. 500 photocolorimeter. The analytical accuracy of these methods ranges from 2% to 5%, and the charge balance errors were generally less than 5%.

The CO_2 and H_2S dissolved gases were determined for 12 thermal springs and wells. The CO_2 was determined according to the method reported in Capasso and Inguaggiato (1998) [41] using a gas chromatograph for analyticcal measurements. H_2S was stabilized with zinc acetate and determined in the laboratory by means of titration.

Selected environmental isotopes were analyzed in 24 of these samples. Stable isotopes of water, ^2H and ^{18}O, and ^{18}O and ^{34}S of dissolved sulfate were determined by mass spectrometry. The standards used were V-SMOV for oxygen and hydrogen and V-CDT for sulfur. ^{18}O of water was determined on CO_2 isotopically equilibratum with H_2O [42], and ^2H was determined from H_2produced by the Zn-reduction method. For ^{34}S analyses, SO_4 was prepared using the methods of Yanagisawa and Sakai (1983) [43], and ^{18}O of sulfate was measured from CO_2 prepared by graphite reduction of $BaSO_4$. All values are reported by delta notation ($\delta‰$). The analytical error is estimated to be less than $\pm 1\%$.

For 15 samples, the tritium concentration was also determined. The samples were distilled, enriched and vacuum-distilled before a liquid scintillation cocktail was added. Analyses were performed using a Perkin-Elmer liquid scintillation counter for twelve 120-min cycles [44]. The results were reported in tritium units (TU).

The air temperature and rainfall data for the area were obtained from the SIMN (Italian hydrographic survey) for the period 1951-1999 [45] and from Regione Lazio for 2000-2010 [46]. The data from the Viterbo meteorological station (327 m asl) were statistically processed to analyze the homogeneity of the data series by applying the cumulative residuals method, and gaps were filled in to complete the series [47,48].

The soil characterization was obtained from the literature [49-51], and the land use information was derived from the Regione Lazio GIS [52].

RESULTS

Hydrostratigraphic Setting

The surface geology grouped by hydrogeological terms is given in Figure 2(a). The sedimentary substratum of the Pleistocene volcanites, together with the likely faults and fractures, are shown inFigure 2(b), as found in the literature [9,53-56] and in the examined lithologic logs.

Representative hydrogeological cross-sections are shown in Figure 3.

The analysis of cross-section and well data highlights a shallow unconfined or leaky aquifer (referred to as the Shallow Aquifer or SA) up to tens of meters thick. The aquifer is contained within the Pleistocene volcanites, which mainly consist of ignimbrites, tuffs and lava flows.

Below the first aquifer, a second confined aquifer (Figure 3) is characterized by thermal waters (referred to as the Thermal Aquifer or TA). This aquifer is intercepted by deeper wells within the volcanites, at the contacts between the volcanites and the flysch units or within the upper portion of the same flysch units, which mainly consist of claystones, marls, marly limestones, sandstones and siliceous limestones.

A low-permeability layer with a thickness of a few meters to tens of meters divides the SA from the TA and is composed of hydrothermally altered pyroclastic deposits or clayey layers of the flysch units.

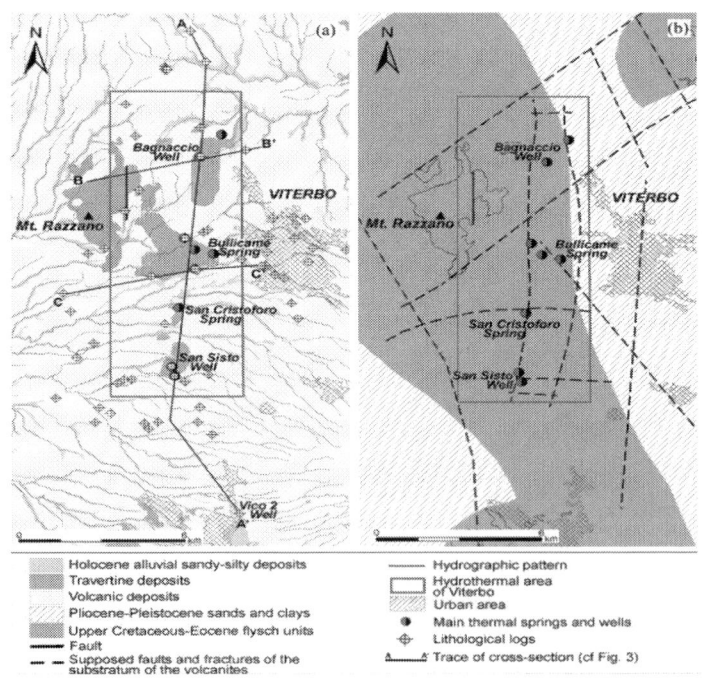

Figure 2: (a) Surface geology of the study area; (b) Sedimentary substratum of the Pleistocene volcanites with the supposed faults and fractures [9,53-56].

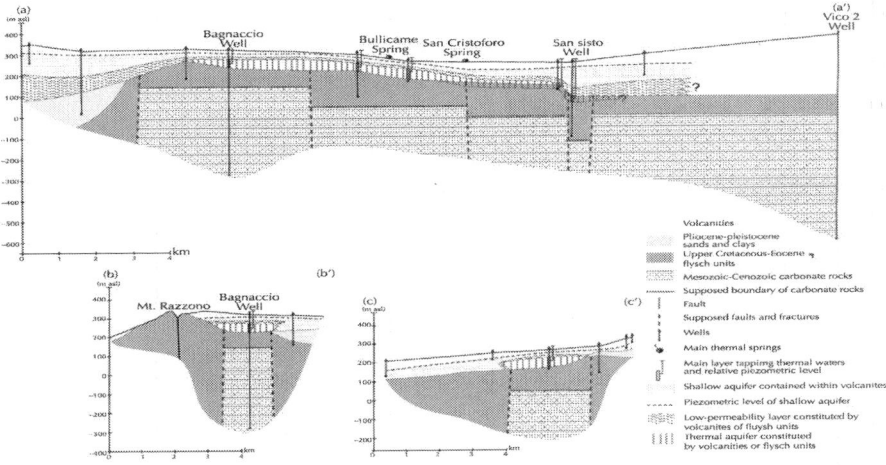

Figure 3: Hydrogeological cross-sections N-S (a-a') and W-E oriented (b-b', c-c'), showing the hydrostratigraphy of the study area.

A third aquifer can be recognized within the deep carbonate rocks, which include limestones, marly limestones, marls, dolomitic limestones, dolostone, and anhydrites. Thermal flow was found in boreholes intercepting this aquifer. In the Bagnaccio Well (Figure 3), the thermal flow was lower than that intercepted in the volcanites and in the flysch units [53,54,57]. In the Vico 2 Well (Figure 3), the thermal flow had the same temperature as that found in the volcanites (62°C) [57]. In Vico 1 Well (V1 in Figure 1(b)), water with a temperature between 50 and 65°C was found in the carbonate rocks [57]. In Vetralla 1 Well (VTR1 in Figure 1(b)), a production test at a depth of 1130 - 1145 m in the carbonate rocks gave a maximum discharge of approximately 15 L/s and a temperature of 61°C [38].

In the Vulsini volcanic area, which is tens of kilometers from the study area, the same Mesozoic-Cenozoic carbonate rocks were recognized as the deep reservoir that feeds two deep geothermal wells characterized by Na(K)-Cl water with a high temperature (120°C - 230°C) and salinity (6 - 12 g/L) [58-63].

The volcanic basement is uplifted, and the flysch units have relatively reduced thicknesses in the hydrothermal area of Viterbo. The SA overlies thick flysch units to the west and low-permeability Pliocene-Pleistocene units to the east, which are mainly constituted by sands and clays. Deep wells east or west of the uplifted block do not tap thermal flow (Figure 3).

Flow and Water Level Measurements

The location of the flow and water level measurements conducted between 2008 and 2010 is shown in Figure 4. Table 1 summarizes the results of the flow measurements.

Thermal water discharges are grouped into zones (Figure 4(b)), and their average values during 2008-2010 are compared with those from 1983-1984 (Table 1).

The cold water discharge is computed from the springs and gaining streams in the area (Table 1and Figure 4(a)). One of the springs, Pidocchio Spring, can be related to the basal water table of the SA if the elevation of the spring (238 m asl) is compared with the piezometric level measured in the neighboring shallow wells; the other springs are related to the perched aquifers of the volcanites. A significant flow arising from the discharge from the basal water table of the SA was measured in the streams of the southern zone during the dry season of 2008 (Figure 4(a)).

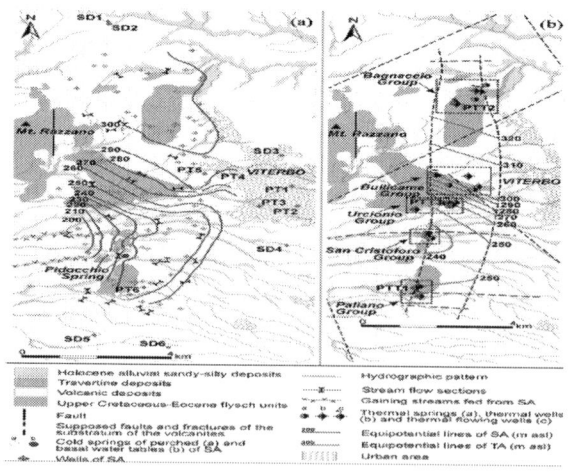

Figure 4: Location of the flow, water level measurements and pumping tests with equipotential maps reconstructed for (a) the shallow aquifer, SA; and (b) Thermal aquifer, TA.

Table 1: Results of flow measurements conducted during 2008-2010 (T: temperature; Q: discharge)

Thermal springs and flowing thermal wells				
Group of springs and wells	**Elevation (nr asp**	**T(°C)**	**Q (L/s) 2008-2010**	**Q (L/S) 1983-1984[a]**
Bagnaccio Group	310 - 320	34 - 62	10.3	17.2
Bullicame Group	284 - 300	40 - 61	29.9	42.3
Urcionio Group	260 - 269	41 - 55	22.8	18.3
S. C'ristoforo Group	225 - 242	35 - 54	2.6	1.8
Paliano Group	240 - 255	54 - 59	13.2	9.0
Total discharge (L s)			78.8	88.6
Springs and increases of streamflow of the shallow aquifer				
Type	Elevation masl)	T(°C)	Q (L/s)	
Spring of basal water-table (Pidocchio Spring)	23S	17	12.5	
Springs of perched water-tables	240 - 284	16 - 32	4.0	
Increase of stream flow	<240		90.7	
Total discharge (L s)			107.2	

From Camponeschi and Nolasco (1984) [57].

The first equipotential map of the SA is based on wells containing water below 23°C. The second map includes wells with temperatures up to 31°C (Figure 4(a)) but is not significantly different from the first map. The equipotential map shows a general conformity of the water-table contours with the topography. The hydraulic gradient varies between 0.006 and 0.06.

A rough potentiometric surface of the TA is given in Figure 4(b). This reconstruction is based on the measurements of the water level or fluid pressure of the deeper wells and the elevation of the springs, which both have water temperatures above 40°C. The map shows two main directions of flow, one oriented NE-SW and a second oriented SE-NW, that converge toward the western boundary of the hydrothermal area.

By comparing the equipotential maps of the TA and SA, a difference in the hydraulic head between 5 and 20 m can be estimated. The values for the vertical gradient are between 0.2 and 1 with a thickness of the low-permeability layer up to 40 m.

Pumping Tests

The wells used for the pumping tests on the SA are reported in Figure 4(a). The test results are given in Table 2.

Table 2: Results of pumping tests of shallow and thermal wells

Shallow wells used for pumping tests				
Well	Q (L/s)	b (m)	Transmissivity (m²/s)	Storativity
PT 1[a]	0.35	26	2.6×10^{-4}	1.2×10^{-3}
PT 2[a]	0.43	17	1.3×10^{-4}	7.8×10^{-3}
PT 3[a]	1.25	34	3.0×10^{-3}	7.9×10^{-3}
PT 4[a]	14.0	14	1.1×10^{-2}	
PT 5[a]	9.7	20	2.3×10^{-2}	1.8×10^{-3}

| PT 5 | 20 | 20 | 4.5×10^{-2} | 6.9×10^{-2} |
| PT 6 | 7.7 | 46 | 5.4×10^{-3} | 5.9×10^{-3} |

Shallow wells used for step-drawdown tests

Well	Q (L/s)	b (m)	Specific capacity s)(m^2/s)	Transmissivity (m^2/s)
SD 1	26	54	2.6×10^{-2}	2.2×10^{-2}
SD 2	25	60	7.8×10^{-3}	5.2×10^{-3}
SD 3	4	40	4.0×10^{-3}	2.4×10^{-3}
SD 4	2	20	1.5×10^{-4}	4.8×10^{-5}
SD 5	2	29	3.3×10^{-4}	1.2×10^{-4}
SD 6	5	21	5.0×10^{-4}	2.0×10^{-4}

Thermal wells used for production tests

Well	Q (L/s)	b (m)	Transmissivity (m^2/s)	Storativity (m^2/s)
PTT 1	38.6	35	8.0×10^{-4} 1.0×10^{-3}	2.0×10^{-4} 3.6×10^{-4}
PTT 2	46.4	55	1.4×10^{-2} 2.8×10^{-2}	
PTT 3	21.5	30	3.9×10^{-3}	

[a]from Piscopo et al. 2006 [37]; Q: discharge; b: saturated thickness.

Five pumping tests (PT1-PT6 in Table 2) were acquired from Piscopo et al. (2006) [37]. The new test conducted on PT5 reveals an increase in temperature from 16.0°C to 16.8°C after 25 hours of pumping and hydraulic parameters comparable to those previously determined. The test conducted on PT6 did not show an increase in temperature or electrical conductivity during pumping.

Six other step-drawdown tests were conducted on the SA (SD1-SD6 in Table 2), and the specific capacity was determined. Using the relationship between transmissivity and specific capacity found for the Cimino-Vico system [34], transmissivity was determined (Table 2).

The tests conducted on wells that intercept thermal waters (PTT1, PTT2 and PTT3) are shown inFigure 4(b), and the results are given in Table 2.

The PTT1 test was conducted at a constant flow for 68 hours on a 125-m-deep well that penetrates fractured flysch formations. The monitored wells and springs are shown in Figure 5(a). The piezometer at the SA and the thermal springs of the San Cristoforo Group did not show significant variations. The San Sisto Well, which is a thermal flowing well, dried up during the pumping and flowed again after the shutdown of the well (Figure 5(b)). The electrical conductivity and temperature of the thermal water were constant in the pumped and observation wells (Figure 5(c)). Other chemical and isotopic parameters did not exhibit significant variations during the pumping period.

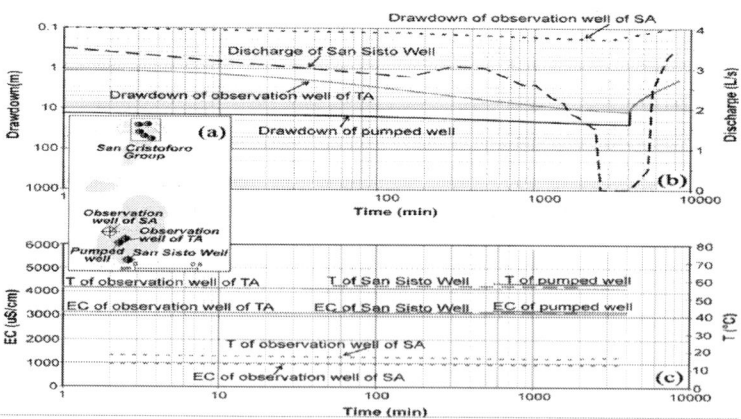

Figure 5: Results of the pumping test conducted on PTT1 well: (a) Location of monitored wells and springs; (b) Trend of drawdown in the tested well and in two observation piezometers, and of discharge of the San Sisto Well;

(c) Trend of temperature (T) and electrical conductivity of waters (EC) of the monitored wells.

The best match of drawdown data from the observation well of the TA was obtained at a transmissivity of 8×10^{-4} to 1×10^{-3} m^2/s and a storativity of 2 to 4×10^{-4} by applying the double porosity model.

The PTT2 test was conducted on the Bagnaccio Well, which was drilled during geothermal exploration during the 1950s [53,54]. The Bagnaccio Well was originally 600 m deep, and it was recently (2008) renovated to a depth of 100 m to better capture the thermal water from the volcanites. The flowing well was tested for 48 hours at a constant rate by measuring the fluid pressure (results in Figure 6). When the flowing well was closed, the pressure immediately returned to its initial value. During the test period, the physical-chemical characteristics of the water and the other monitored chemical and isotopic parameters did not change. Among the wells and springs monitored during the pumping, only the Bagnaccio Spring (78 m from the well) showed a significant variation (Figure 6).

An approximate transmissivity of 2.8×10^{-2} m^2/s was estimated by applying the Cooper-Jacob method to the drawdown data measured in the production well. The distance-drawdown method, which was applied considering the Bagnaccio Spring as a piezometer, permitted to determine transmissivity values between 1.4×10^{-2} and 2.3×10^{-2} m^2/s (Table 2).

A third test on the thermal aquifer was conducted in the central zone (PTT3 test) by closing a well that normally flows six days a week at a constant rate of 21.5 L/s. Recovery was observed at a second well 129 m away. The two wells are 42 and 93 m deep and capture thermal water from the volcanites. The transmissivity was estimated to be 3.9×10^{-3} m^2/s when considering the residual drawdown in the second well, the times since the initiation and termination of pumping, and the discharge of the flowing well.

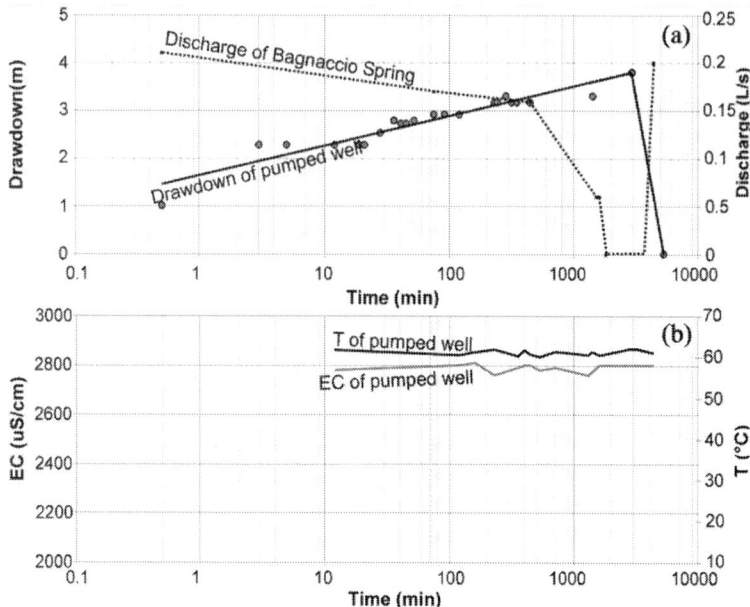

Figure 6: Results of the pumping test conducted on PTT2 well: (a) Trend of drawdown in the tested well and of discharge of the Bagnaccio Spring; (b) Trend of temperature (T) and electrical conductivity of waters (EC) of the tested well.

General Chemistry

The main chemical constituents of all of the samples are provided in Table 3. The samples were categorized as one of the following: thermal water (spring, ts, and well, tw) from TA; spring (s) and well (w) of the SA; and stream waters (st). The location of the sampled water is shown in Figure 7.

According to the physical-chemical characteristics and the relative abundance of major cations and anions, the thermal waters (also including the waters of Bagnaccio Pond, 7 in Table 3 and Figure 7, which are influenced by the flow of the SA) exhibit a homogenous hydrochemical facies. They are calcium-sulfate waters (Figure 8) that are characterized by a temperature (T) of 40°C to 62°C, a pH less than 7, and a specific electrical conductivity (EC) between 2800 and 3600 µS/cm.

Wells and springs fed from the SA show a more heterogeneous hydrochemical facies and range from calciumalkaline-bicarbonate to calcium-sulfate waters (Figure 8), with a T of 16°C to 31°C, pH up to 8, and EC of 300 to 2900 μS/cm. The two sampled streams have opposite geochemical profiles: one is recharged by the SA (i.e., 15 in Table 3 and Figure 7), whereas the other also by the TA (i.e., 32 in Table 3 and Figure 7).

Among the minor and trace constituents, Li (generally between 0.01 and 0.1 mg/L) and Fe (generally between 0.01 and 0.2 mg/L) are present at very low concentrations. The NO_3^-concentration is lower in thermal waters (generally less than 5 mg/L) than in the waters of the SA (up to 100 mg/L, relative to the depth of the water level below the ground). The SiO_2 concentration varies between 40 and 55 mg/L for thermal waters and is generally lower for the other waters. The strontium concentration is higher in thermal waters, as is the fluoride concentration (Table 3).

The Pearson correlation matrix of the entire dataset shows a strong correlation among EC, T, Mg^{2+}, S_4^{2-}, Sr, and HCO_3^- (R > 0.8). The plot of Sr versus the sulfate concentration in Figure 9highlights one of these correlaons.

The concentrations of dissolved CO_2 and H_2S gas in the 12 thermal waters (4, 8, 9, 14, 16, 18, 20, 30, 34, 44, 45, 48 in Figure 7) vary between 300 and 600 mg/L and 7 and 30 mg/L, respectively.

The highest values were found in the deeper wells that had proper screening, casing and sealing.

These results agree with the previous hydrochemical characterization [37,40] and further highlight that the geochemical profile of the waters of the SA appears to be influenced by mixing with thermal waters (Figures 8 and 9). To analyze this point thoroughly, sulfate content was considered as one of the constituents that discriminates the different waters (see also the stable isotope analysis). The following relationship, which has been used in the literature to determine the components of stream flow [64], was applied:

$$Ci \cdot Qi = Ct \cdot Qt + Cc \cdot Qc$$

(1)

where Ci is the sulfate concentration in water sampled from the SA; Ct is the average sulfate concentration of the thermal waters (1193 mg/L) and is one of the end members;

Cc is the sulfate concentration in Pidocchio Spring (39 in Table 3 and Figure 7), which represents the other end member (19 mg/L), i.e., cold water from the SA that is not influenced by the thermal flow;

Qi is the total flow rate of the SA;

Qt is the component of the flow rate of thermal waters in the SA; and Qc is the component of the flow rate of cold waters in the SA.

Considering that Qi = Qt + Qc, Equation (1) can be rewritten as

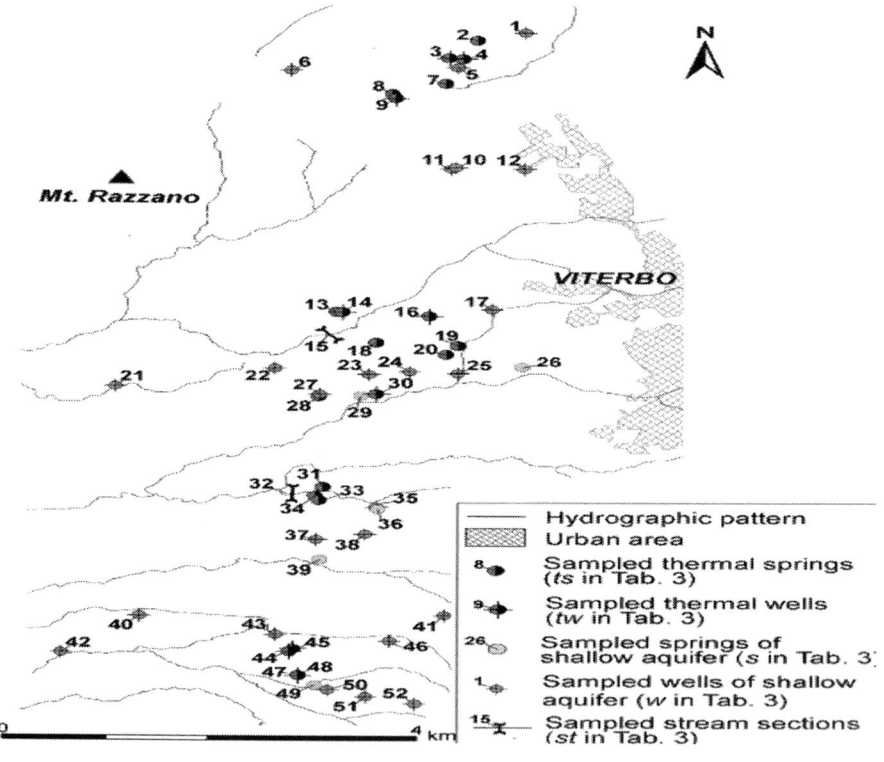

Figure 7: Location of sampled waters (IDs in Table 3).

Table 3: Main chemical constituents of sampled waters (June 2009)

ID	Type	T (°C)	pH	EC (μS/cm)	Na+ (mg/L)	K+ (mg/L)	Ca2+ (mg/L)	Mg2+ (mg/L)	Cl- (mg/L)	HCO3- (mg/L)	SO4 (mg/L)	SiO3 (mg/L)	Sr (mg/L)	F (mg/L)
1	w	24.0	7.31	1450	29	19	205	78	21.8	463	465	41.0	2.6	1.9
2 (Ba)	ts	50.0	6.70	3011	36	29	515	145	13.0	976	1100	43.3	11.4	4.0
3 (Ba)	tw	62.4	6.85	3290	49	30	504	150	16.2	1055	1155	47.5	11.6	3.9
4 (Ba)	tw	59.6	6.57	3300	36	29	508	145	17.8	1094	1140	47.3	11.4	4.2
5	w	17.0	7.40	1630	27	20	230	60	27.2	610	545	39.6	4.7	1.9
6	w	18.6	7.72	738	44	35	81	24	78.3	249	58	37.5	0.47	0.5
7 (Ba)	ts	33.6	6.20	3360	44	30	542	150	18.2	945	1300	40.3	12.2	3.2
8 (Ba)	ts	62.3	6.30	3170	35	28	487	150	14.4	985	1106	18.2	11.7	3.6
9 (Ba)	tw	61.7	6.32	2780	33	26	527	165	15.4	-	1270	40.3	11.3	4.7
10	w	29.4	6.10	2920	38.	28	468	115	23.8	1073	1023	29.6	9.7	1.5
11	w	25.2	6.20	2890	41	28	471	115	21.0	1061	956	40.0	9.9	1.3
12	w	17.0	7.80	535	41	16	43	12	19.0	189	53	39.3	2.2	3.6
13 (Bu)	ts	58.9	6.41	3230	34	29	500	141	15.2	1005	1115	45.0	12.1	4.2
14 (Bu)	tw	60.6	6.44	3200	34.	30	502	140	15.6	1018	1100	49.1	11.7	4.3
15	st	19.9	8.07	1255	44	21	168	40	49.4	457	282	39.4	2.7	1.9
16 (Bu)	tw	54.5	6.43	2900	34	31	456	115	17.6	1030	976	47.1	10.4	4.0
17	w	21.6	7.80	747	23	14	126	11	20.2	305	129	38.9	2.9	3.8
18 (Bu)	ts	57.8	6.40	3000	34	33	510	120	15.1	1067	986	48.2	11.1	3.3
19 (Bu)	tw	40.0	6.37	2770	34	27	496	109	17.0	976	946	45.2	9.5	2.6
20 (Bu)	ts	54.8	6.27	2990	35	36	500	115	12.3	1006	1091	53.6	11.1	4.0
21	w	21.3	7.71	756	46	26	73	14	21.6	177	141	40.2	0.60	4.0
22	w	21.6	6.93	1950	18	10	318	59	15.6	601	548	31.1	4.2	3.1
23	w	30.5	6.63	2880	30	24	507	112	17.0	1036	971	38.4	10.7	5.2
24	w	24.7	7.47	1697	27	34	225	110	14.0	378	650	39.4	6.9	2.4
25	w	20.0	7.40	1580	32	25	205	75	23.2	451	513	40.3	5.2	2.2
26	s	19.6	6.80	725	35	34	71	18	12.7	378	68	41.3	0.27	1.1
27	w	29.6	6.64	2480	38	16	465	55	48.0	754	912	43.3	5.6	3.6
28 (Ur)	tw	41.5	6.29	3490	67	28	588	140	20.4	1125	1299	41.3	12.2	4.4
29	s	31.4	6.63	2040	50	20	336	83	27.8	724	704	22.3	5.6	1.8
30 (Ur)	tw	52.5	6.37	3030	46	26	504	125	16.4	1021	1111	18.2	11.7	2.6
31 (Sc)	ts	53.6	6.40	3570	34	24	635	165	19.0	1066	1400	40.9	13.9	5.2
32	st	27.0	7.32	1930	34	22	302	72	21.4	549	769	38.9	5.5	3.2
33 (Sc)	ts	52.5	6.20	3520	31	26	618	150	15.6	1069	1401	51.3	13.3	3.6
34 (Sc)	ts	43.1	6.75	3560	30	26	584	150	15.0	1024	1346	44.2	13.7	3.1
35	w	21.1	7.45	393	23	21	43	10	28.2	140	42	36.6	0.34	1.8
36	s	16.0	7.77	490	28	10	36	11	29.1	116	46	39.5	0.54	0.5
37	w	25.0	8.37	350	17	18	27	8	15.6	132	18	34.5	0.17	0.8
38	w	21.3	7.80	532	25	20	46	11	42.7	52	57	37.6	0.38	1.1
39	s	17.0	7.86	327	25	22	15	8	23.1	101	19	31.3	0.17	0.8
40	w	20.8	6.64	1330	40	16	205	41	35.7	483	342	40.6	2.8	4.2
41	w	21.0	7.66	356	24	19	27	6	25.7	97	26	33.6	0.15	2.4
42	w	20.5	6.55	2360	27	24	376	90	21.0	729	952	42.3	8.0	4.0
43	w	20.2	7.88	880	9.3	4.6	157	10	12.0	354	69	42.0	1.3	1.7
44 (Pa)	tw	59.5	6.42	3540	43	26	637	155	23.2	1055	1405	39.8	13.2	3.4
45 (Pa)	tw	58.5	-	3230	32	26	632	170	19.6	-	1300	44.0	13.2	3.2
46	w	25.0	6.78	930	25	18	158	19	21.6	418	140	37.5	2.0	3.0
47	s	24.5	6.61	713	26	18	91	14	22.0	201	134	36.6	0.81	2.6
48 (Pa)	tw	58.2	6.39	3510	47	26	617	137	16.0	1069	1311	46.5	13.1	4.5
49	s	19.2	7.35	328	22	22	28	6	20.6	104	26	32.4	0.14	2.2
50	w	23.0	7.37	332	26	22	30	6	23.0	94	31	30.6	0.13	2.3
51	w	21.2	7.70	368	27	21	23	5	24.2	91	51	31.0	0.11	3.2
52	w	20.2	7.60	367	28	20	17	6	25.4	90	33	37.0	0.13	2.9

Ba: Bagnaccio Group; Bu: Bullicame Group; Ur: Urcionio Group; Sc: San Cristoforo Group; Pa: Paliano Group; ts: thermal spring; tw: thermal well; s: spring of the shallow aquifer; w: well of the shallow aquifer; st: stream section

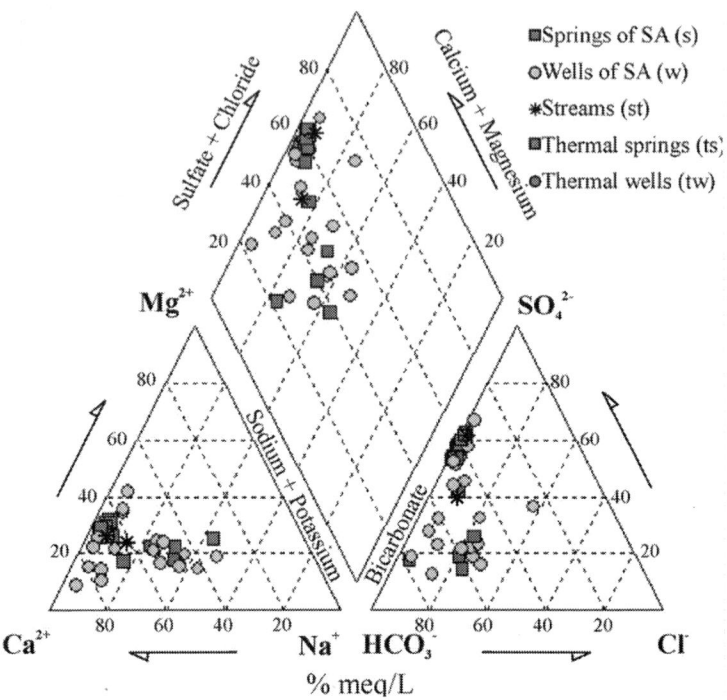

Figure 8: Piper diagram of sampled waters.

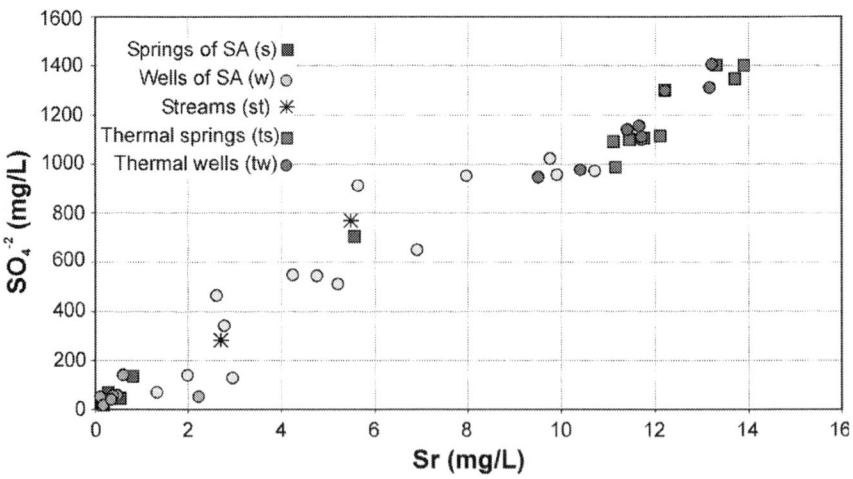

Figure 9: Plot of strontium versus sulfate contents of sampled waters.

$$\frac{Qt}{Qi} = \frac{Ci - Cc}{Ct - Cc}$$

The ratio Qt/Qi permits us to determine the fraction of thermal water in the groundwater of the SA. As reported in Figure 10(a), Qt/Qi is between 0.1 and 0.5 in the central and northern zones of the Viterbo hydrothermal area and between 0.1 and 0.6 in the western zone beyond the boundary of the hydrothermal area. The same distribution of the Qt/Qi ratio is found when strontium is used as an indicator of the mixing between the thermal waters and cold waters of the SA (Figure 10(b)).

Stable Isotopes and Tritium

The stable isotopes of water ($d^{18}O$ and d^2H), which were determined for thermal waters (2, 3, 4, 8, 9, 14, 16, 18, 19, 20, 30, 34, 44, 45, 48 in Figure 7), selected wells (6, 21, 23, 41 in Figure 7) and springs of the SA (26, 29, 39 in Figure 7) and two stream waters (15, 32 in Figure 7), are plotted in Figure 11. In the same graph, the global [65] and central-Italy meteoric water lines [66] and the isotopic contents of Vico Lake [37] are reported. All of the waters exhibit $d^{18}O$ and d^2H values in a limited range (–5.6‰ to –7.2‰ and –34‰ to –44‰, respectively) and fall on the meteoric water lines, as determined by previous studies [37]. The waters with lower $d^{18}O$ values are from the TA and the basal water table of the SA. If the available vertical isotopic gradients for the western side of the Apennine Chain [66] are considered, then the elevation of the recharge area of the sampled waters ranges between 330 and 1270 m·asl. When only the waters that are more enriched in $d^{18}O$, i.e., mainly those of the perched aquifers of the SA, are considered, the elevation is less than 470 m·asl.

The results for the stable isotopes of dissolved sulfate ($d^{34}S_{SO4}$ and $d^{18}O_{SO4}$) are plotted in Figure 12, which was modified from Clark and Fritz (1997) [67]. For the thermal waters, the $d^{34}S_{SO4}$ values vary from 11.4‰ to 16.8‰, and the $d^{18}O_{SO4}$ values vary from 11.3‰ and 14.2‰, with generally lower values for the SA and stream waters. According to the diagram of Clark and Fritz (1997) [67], the thermal waters plot within the Devonian to lower Triassic rectangle, in contrast to the SA and stream waters, which seem to be related to the atmospheric content or mixing with the two previous components. The Mesozoic-Cenozoic carbonate rocks, which constitute the deep substratum of the area under investigation, include Triassic anhydrites.

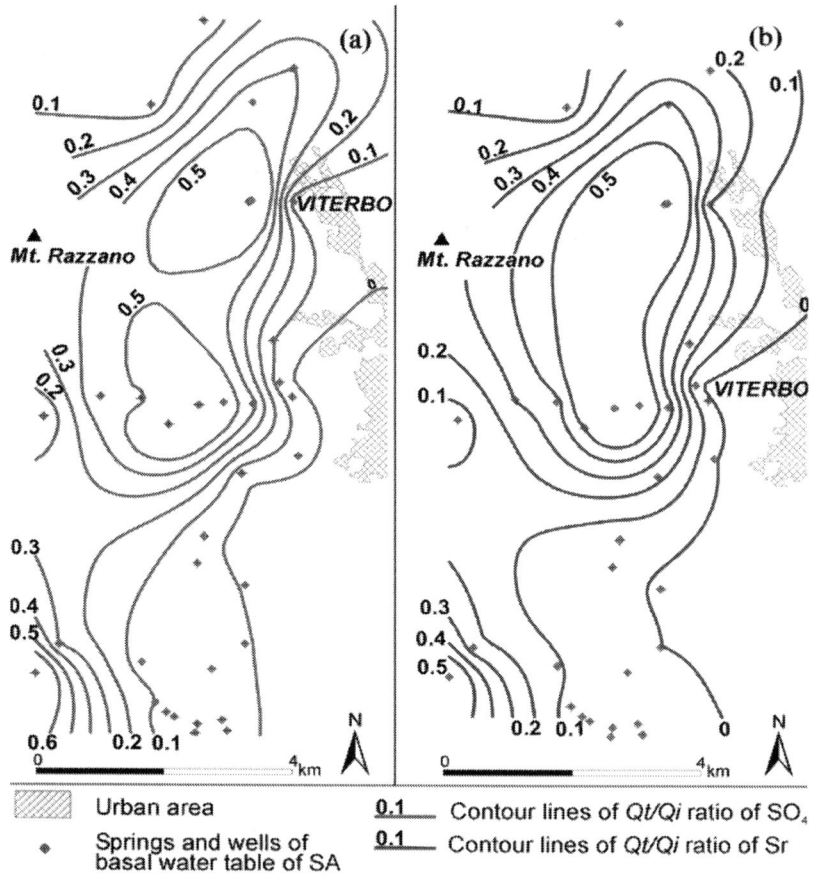

Figure 10: Map of the ratio Qt/Qi which represents the fraction of thermal water (Qt) in the groundwater of the SA (Qi), using (a) sulfate concentration and (b) strontium concentration.

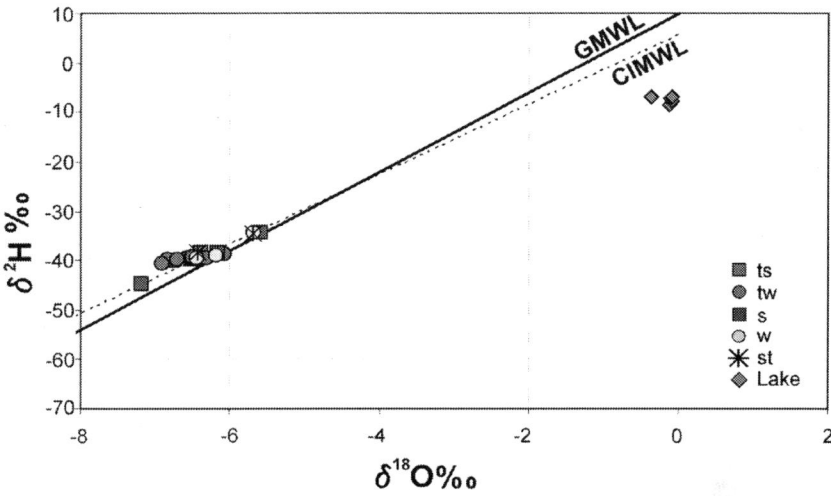

Figure 11: d^{18}O versus d^{2}H in thermal waters (ts and tw), some wells (w) and springs (s) of SA and stream waters (st), compared with isotopic contents of waters of Vico Lake [37]. The global meteoric (GMWL) and central Italy meteoric water lines (CIMWL) are also shown.

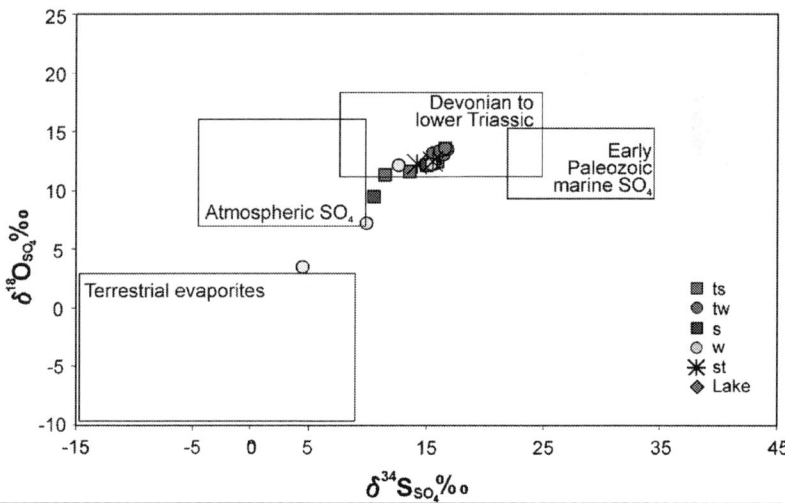

Figure 12: d^{34}S$_{SO4}$ versus d^{18}O$_{SO4\ in\ dissolved\ sulfate\ of\ ther}$- mal waters (ts and tw), some wells (w) and springs (s) of SA and stream waters (st) (fields from Clark and Fritz, 1997 [67]).

As can be seen from the plot of Figure 12, some SA samples fall in the Devonian to lower Triassic rectangle, which again confirms the mixing with thermal waters.

Of the samples, 15 were analyzed for tritium, of which 12 were from thermal waters (2, 4, 8, 9, 14, 16, 18, 20, 30, 34, 45, 48 in Figure 7) and 3 were from cold waters (6, 21, 41 in Figure 7). The tritium concentrations varied from 2 to 11 TU, which suggests a recent component recharge of waters, i.e., post-1952 [67]. The thermal waters had lower tritium concentrations (2 to 5 TU), whereas samples from wells exclusively intercepting the SA had higher concentrations (8 to 11 TU). If the radioactive half-life for tritium is considered and the same isotopic content of rainwater recharging is assumed for both thermal and cold waters, then there is a 14-year-difference in the residence time between the two types of waters.

Soil-Water Budget

The soil-water budget was estimated with reference to a surface area of approximately 38 km^2, including the strip where thermal waters flow out.

The mean annual values of precipitation and temperature for 1989 to 2010 were calculated by processing the meteorological data recorded in Viterbo. The homogeneity of the datasets was evaluated, and missing data were reconstructed on a monthly basis (2% of the precipitation data and 0.25% of the temperature data were missing). The mean annual values for precipitation and temperature are 772 mm and 15.7°C, respectively.

Based on the mean monthly rainfall and temperature data, a potential evapotranspiration of 841 mm/y was calculated using the Thornthwaite empirical formula [68]. The actual evapotranspiration was also determined by the Thornthwaite-Mather method [69] by considering the soil texture, field capacity, permanent wilting point and land cover for the different zones in the area. The resulting total available water-holding capacity varies between 80 and 160 mm; therefore, the actual evapotranspiration varies between 525 and 605 mm/y.

The mean annual actual evapotranspiration is 575 mm for the strip where thermal waters flow out (surface area of 19.66 km^2), and the difference between the mean annual precipitation and actual evapotranspiration is 197 mm. For the northern and central zones of Figure 10 with Qt/Qi between 0.1 and 0.5 (surface area of 18.79 km^2), the mean annual actual evapotranspiration is 581 mm, and the difference between the mean annual precipitation and actual evapotranspiration is 191 mm.

DISCUSSION

A refinement of the conceptual model and a groundwater resource estimation of the Viterbo thermal area can be derived from the combination of hydrogeological and hydrochemical data.

The hydrogeological interpretation of the stratigraphy enables the characterization of the upper 100 - 200 m of the two main aquifers. The shallow aquifer (SA) consists of Pleistocene volcanites and covers the entire study area; the deeper aquifer (TA) is characterized by thermal waters. Within the study area, the two aquifers are generally separated by a low-permeability layer of volcanites or flysch units. At greater depths, a thick low-permeability layer consisting of flysch units is locally fractured and faulted and overlaps the deep carbonate rocks that are also faulted and dislocated (Figure 13).

The unconfined or leaky SA has a thickness of a few meters to tens of meters. The SA exhibits a piezometric surface that is consistent with that of the wide CiminoVico aquifer system. In the study area, the SA is recharged by direct infiltration and groundwater inflow from the Cimino-Vico system, discharges locally in streams and in springs, and westward groundwater outflow occurs. The hydraulic parameters of the SA consist of transmissivity values between 10^{-5} m^2/s and 10^{-2} m^2/s and storativity values between 10^{-3} and 10^{-2}.

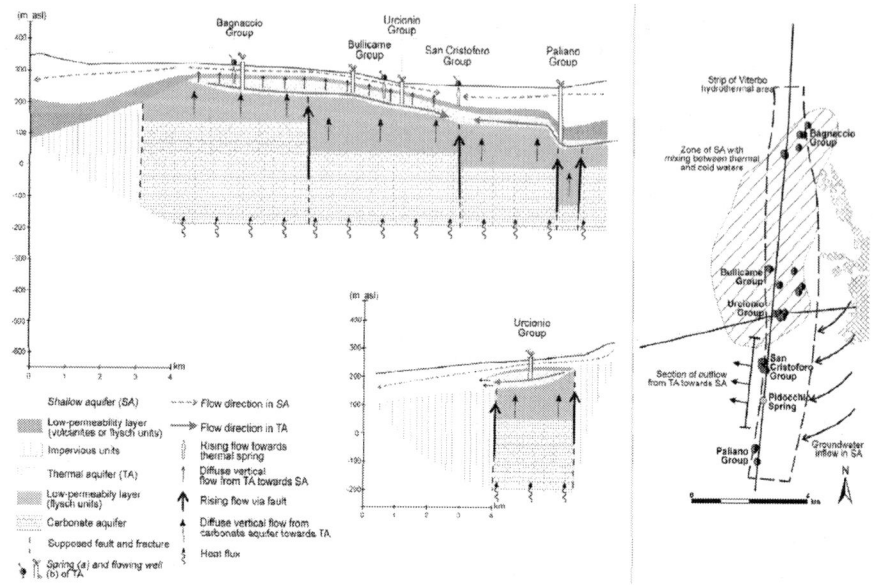

Figure 13: Hydrogeological conceptual model of the hydrothermal area of Viterbo showing the main groundwater paths in and among the overlapping aquifers. Zones considered for the evaluations of the yield of the groundwater system are shown on the right.

Waters circulating in the SA are generally characterized by low temperature (less than 23°C) and salinity (EC less than 700 µS/cm) and are of bicarbonate-alkalineearth or bicarbonate-alkaline types, which suggests a short duration of the rock-water interaction. Sampled waters from the SA included a hydrochemical facies that arose from mixing between typical cold waters and thermal waters. These waters have been found in the northern and central zones of the study area, where the SA overlies the TA, and in the western boundary of the hydrothermal area, where the TA has not been identified. In these zones, the flow from the TA influences 10% to 60% of the flow in the SA, based on sulfate and strontium concentrations.

The TA is located at a depth of up to 200 m and is characterized by thermal waters (T of 40°C to 62°C) with higher salinities (EC from 2800 to 3600 µS/cm). The thickness of the volcanites and flysch units constituting the TA varies between 50 and 80 m, even if other layers with thermal waters have been found in the same deep flysch units [53, 54].

The TA has been identified within a rectangular area that is approximately 12 km long and 2 km wide (Figure 13), and it includes the following characteristics: 1) an uplift of sedimentary units underlying the volcanites; 2) a limited thickness of the same volcanic cover as the surroundings; 3) a fractured and faulted zone of the sedimentary basement; 4) a zone with one of the higher geothermal gradients in the Latium region (up to 100°C /km) and, therefore, a high heat flow (from 100 to 400 mW/m^2) [1,26]; and 5) an outcropping of thermogene travertine and CO_2 emissions [33]. Within this N-S elongated zone, the TA is continuous (Figure 13). The southern boundary of the TA is not certain because the thermal waters have been intercepted by the Vico 2 Well in the volcanites (at a depth of 275 to 290 m) [57] and, recently, in the flysch units 10 km south of the area under investigation.

The confined TA is characterized by two main directions of flow that converge westwards, and its hydraulic parameters consist of transmissivity values of 10^{-2} m^2/s to 10^{-4} m^2/s and a storativity of approximately 10^{-4}. The hydraulic tests highlight the continuity of the TA and are in agreement with the results of production tests conducted in the 1950s [53, 54].

The TA discharges into thermal springs and flowing wells. A diffuse vertical flow from the TA toward the SA through the aquitard also occurs, particularly in the northern and central zones of the strip where the two aquifers overlap, according to the vertical gradient between the two aquifers and the mixing highlighted by the chemical and isotopic data. Moreover, an outflow from the TA towards the SA occurs at its western boundary (Figure 13): 1) hydrostratigraphy highlights the lateral connection between the TA and the SA; 2) waters sampled from shallow wells show mixing with thermal waters; 3) the hydraulic heads of the TA and the SA are not as different. The northern and eastern boundaries of the TA do not present the full set of features.

Waters circulating in the TA are characterized by calcium-sulfate hydrochemical facies, a high dissolved gas content (CO_2 and H_2S), high temperature and salinity. These features, together with those obtained from the analyses of the minor and trace constituents (such as Sr and F$^-$), suggest deeper circuits and longer rock-water interactions compared with those occurring in the SA. The stability of the water quality based on the pumping tests and the chemical homogeneity of the thermal waters confirms the continuity of the TA. Isotopic analyses strengthen this hydrogeological conceptual model. The stable isotopes of water highlight the meteoric origin of the thermal waters and that their recharge area is similar to that of the SA. The waters of the two aquifers differ in residence time; the TA waters have an isotopic age of less than 50 - 60 years, which is approximately ten years older than the cold waters. Other chemical and isotopic indicators, such as the high sulfate and strontium content and $d^{34}S_{SO4}$ and $d^{18}O_{SO4}$ values, can be explained through interactions with the fluid circulating in the deep carbonate basement.

The deep carbonates were recognized as the main reservoir of hot fluids feeding geothermal wells in the Vulsini volcanic area e.g., [60, 63], which is tens of kilometers from the study area. Documented evidence from four deep wells in and surrounding the area under examination confirms groundwater circulation in the carbonate aquifer [38, 53, 54, 57]. Therefore, the deep carbonate aquifer can be considered to be a reservoir that recharges the TA, with a flow that mainly rises via faults and fractures in the sedimentary substratum of the volcanites because of the high heat flow that characterizes the region.

Considering the limited range of $d^{18}O$ and d^2H in all waters sampled and the slight difference in tritium content between the thermal waters and those of the SA, the recharge area of the TA appears to be the same as that of the SA, i.e., the Lake Vico area of the Cimini Mountains. According to Piscopo et al. (2006) [37], this circuit seems to be consistent with the difference between the hydraulic head of the recharge area (approximately 500 m·asl) and that determined for the thermal waters (from 225 and 320 m·asl).

Based on previous conceptual models, the yield of the groundwater system was estimated with reference to the strip (Figure 13) in which the thermal waters flow out (surface area of 19.66 km²).

The total thermal water discharge (Q_{tw}) can be written as

$$Q_{tw} = Q_{sw} + Q_{vt} + Q_{lo}$$

(2)

where Q_{sw} is the discharge into thermal springs and flowing wells;

Q_{vt} is the vertical flow from the TA towards the SA, which occurs in the northern and central zones of the strip; and Q_{lo} is the flow from the TA towards the SA, which occurs mainly in the western boundary of the strip.

Q_{sw} is approximately 84 L/s, considering the average discharge measured in 1983-84 and 2008-10 (Table 1).

Q_{vt} can be evaluated by considering the direct infiltration in the northern and central zones of the SA (I_{nc}), where mixing between thermal and cold waters has been found (Figure 13), and the ratio of the flows of the two end members derived from Equation (1), rewritten as Qt/Qc:

$$Q_{vt} = I_{nc} \cdot \frac{Qt}{Qc}$$

I_{nc} was estimated to be approximately 91 L/s, taking into account the following: 1) the surface area where mixing has been found (18.79 km²); 2) the mean annual difference between the precipitation and actual evapotranspiration calculated for the northern and central zones (191 mm); and 3) an I_{nc} equal to 0.8 of the difference between the precipitation and the actual evapotranspiration that is attributed to the flat topography of the area and the absence of streamflow for most of the year. The average Qt/Qc for the area under examination is 0.75, which gives a value of 68 L/s for Q_{vt}.

Q_{lo} can be evaluated by considering Qt/Qi for the western zone of the SA beyond the boundary of the strip and the horizontal flow in the SA, which corresponds to Qi, by applying Darcy's Law for the section of outflow from the TA toward the SA (Figure 13). Q_{lo} is approximately 53 L/s because the flow in the SA is approximately 220 L/s (if the mean transmissivity of this zone is considered, i.e., the mean obtained from the PT6, SD5 and SD6 tests in Table 2), and the average Qt/Qi is 0.24.

The total discharge of the thermal waters, Q_{tw} is equal to 205 L/s.

The potential yield of the TA was independently estimated, taking into account the potentiometric surface and transmissivity values determined for the TA. Two sections of flow were considered according to the main flow directions. The first section is in the northern zone, and the second is in the southern zone of the strip; both sections were chosen to be distant from thermal springs and flowing wells. By assigning the lower values obtained from the pumping tests of the northern and southern sections and the proper values of the hydraulic gradient, a total flow rate of approximately 250 L/s results from the application of Darcy's Law.

Although these evaluations may be subject to uncertainty because of the methods employed and the uncertainty in the data, it is clear that the yield of the TA is higher than what is discharged into the springs and wells. The TA can be considered to have a minimum yield of approximately 170 L/s in the examined strip, even with a 30% error in Equation (2) (i.e., Q_{vt} and Q_{lo}).

To complete the evaluation of the yield of the groundwater system, the mean potential direct recharge was estimated for the SA in the strip under examination. The result was approximately 98 L/s, which considers the mean annual difference between the precipitation and the actual evapotranspiration of the entire strip. The yield of the SA in the strip is higher than the direct recharge because of the groundwater inflow from the Cimino-Vico system in the eastern boundary of the strip. In the strip, the effective irrigation consumption from the SA can be evaluated by taking into account the difference between the potential (841 mm) and actual evapotranspiration (577 mm) of the irrigated zones included in the strip (12.46 km^2) and considering that the over-irrigation returns to the shallow aquifer. The estimated effective irrigation consumption is 104 L/s, which is comparable to the direct recharge.

CONCLUSIONS

The hydrogeological conceptual model seems to be more complex than the models presented previously in the literature. The shallowest 100 - 200 m of the two main aquifers, which are mainly tapped for drinking water, irrigation and spas, have been characterized. At a greater depth, another aquifer was recognized as the reservoir of the hot waters feeding the shallower waters. The hydraulic and chemical characteristics of the two shallower aquifers that were examined in detail in this study are very different. The first aquifer (SA) is related to the regional groundwater flow of the Cimino-Vico system, is hydraulically heterogeneous and is generally characterized by cold, fresh waters. The second aquifer (TA) is continuous within a strip in which the local hydrostratigraphic, structural and geothermal conditions allow a relatively active and constantly replenished flow of thermal waters with higher salinity. These two aquifers interact vertically and laterally, to give rise to mixed waters circulating in the first aquifer.

The yield of the groundwater system was estimated by an integrated hydrogeological and hydrochemical approach. In the examined strip, the yield of the TA is at least 170 L/s, and it discharges into thermal springs and wells and feeds the SA vertically and laterally. The SA is recharged by direct infiltration and inflow from regional groundwater as well as inflow from the second aquifer.

Based on the conceptual hydrogeological model and the previous estimate of the yield of the groundwater system, some preliminary considerations regarding groundwater management can be made.

Even if, at the scale of the whole groundwater flow system, the potential exists for future development of the TA, then an increase in withdrawals through wells from the TA could result in a decrease in the discharge from the thermal springs, which has occurred in the past for some thermal springs and was verified during the pumping tests. Therefore, the future use of thermal waters must account for the potential yield of the TA and provide sufficient residual discharge at thermal springs for recreational use.

The effects of withdrawals from the SA on the entire groundwater system may also be important if it is considered that the flow rate necessary for irrigation in the strip is comparable to the direct recharge in the area. At the local scale, pumping from the SA could increase the vertical gradient between the two overlapping aquifers to cause an increase in flow from the TA toward the SA. This could cause an increase in the temperature and salinity of the SA that is tapped for irrigation and drinking water as well as a decrease in the flow rate of the thermal waters tapped to supply spas and for recreational use.

These examples highlight that, in the volcanic hydrogeological environment, one of the most stringent constraint in determining the correct usage of a resource is the co-existence of interacting groundwater flows of different qualities. However, in this fragile and complex hydrogeological system with a multi-purpose water demand, future decisions regarding groundwater management must also be based on economic, legal and environmental criteria to define the priorities for the use of different groundwater resources.

ACKNOWLEDGEMENTS

The authors are thankful to the Regione Lazio—Direzione Regionale Attività Produttive e Rifiuti, in the persons of Dr Mario Marotta, Eng Luigi Minicillo and Dr Patrizia Refrigeri, and Ministero dell'Istruzione, dell'Università e della Ricerca (Project PRIN-2008 - 2008YYZKEE_02) for financial support.

REFERENCES

1.   R. Cataldi, M. Mongelli, P. Squarci, L. Taffi, G. Zito and C. Calore, "Geothermal Ranking of Italian Territory," Geothermics, Vol. 24, No. 1, 1995, pp. 115-129. doi:10.1016/0375-6505(94)00026-9.

2.   G. Bertrand, H. Celle-Jeanton, F. Huneau, S. Loock and C. Renac, "Identification of Different Groundwater Flowpaths within Volcanic Aquifers Using Natural Tracers for the Evaluation of the Influence of Lava Flows Morphology (Arghat Basin, Chaîne des Puys, France)," Journal of Hydrology, Vol. 391, No. 3-4, 2010, pp. 223-234.doi:10.1016/j.jhydrol.2010.07.021

3.   J. V. Cruz and M. O. Silva, "Hydrogeologic Framework of the Pico Island (Azores, Portugal)," Hydrogeology Journal, Vol. 9, No. 2, 2001, pp. 177-189.doi:10.1007/s100400000106.

4.   E. Custodio, "Groundwater Characteristics and Problems in Volcanic Rock Terrains," In: IAEA, Isotope Techniques on the Study of the Hydrology of Fractured and Fissured Rocks, STI/PUB 790, Vienna, 1989, pp. 87-137.

5.   E. Custodio, "Groundwater in Volcanic Rocks," In: Krasny, J. and J. M. Sharp, Eds., Groundwater in Fractured Rocks, Taylor and Francis Group, London, 2007

6.   V. Piscopo, V. Allocca and F. Formica, "Sustainable Management of Groundwater in Neapolitan Volcanic Areas, Italy," In: O. Sililo, et al., Eds., Groundwater: Past Achievements and Future Challenge, Balkema, Rotterdam, 2000.

7.   V. Piscopo, A. Baiocchi, S. Bicorgna and F. Lotti, "Hydrogeological Support for Estimation of the Sustainable Well Yield in Volcanic Rocks: Some Examples from Central and Southern Italy," Proceedings of 36th IAH Congress on Integrating Groundwater

Science and Human Well-Ceing, Toyama, 26 October-1 November 2008, pp. 1652-1666.

8. P. Ambrosetti, M. G. Carboni, M. A. Conti, A. Costantini, U. Esu Nicosia, G. Parisi and F. Sandrelli, "Evoluzione paleogeografica e tettonica dei bacini tosco-umbro laziali nel Pliocene e nel Pleistocene inferiore," Memorie della Società Geologica Italiana, Vol. 19, 1978, pp. 573-580.

9. P. Baldi, F. A. Decandia, A. Lazzarotto and A. Calamai, "Studio Geologico del Substrato della Copertura Vulcanica Laziale Nella Zona dei Laghi di Bolsena, Vico e Bracciano," Memorie Società Geologica Italiana, Vol. 13, No. 4, 1974, pp. 575-606.

10. F. Barberi, G. Buonasorte, R. Cioni, A. Fiordelisi, L. Foresi, S. Iaccarino, M. A. Laurenzi, A. Sbrana, L. Vernia and I. M. Villa, "Plio-Pleistocene Geological Evolution of the Geothermal Area of Tuscany and Latium," Memorie Descrittive della Carta Geologica d'Italia, Vol. 49, 1994, pp. 77-134.

11. G. Marinelli, F. Barberi and R. Cioni, "Sollevamenti Neogenici e intrusioni acide della Toscana e del Lazio settentrionale," Memorie della Società Geologica Italiana, Vol. 49, 1993, pp. 279-288.

12. L. Beccaluva, P. Di Girolamo and G. Serri, "Petrogenesis and Tectonic Setting of the Roman Volcanic Province, Italy," Lithos, Vol. 26, No. 3-4, 1991, pp. 191-221.doi:10.1016/0024-4937(91)90029-K

13. S. Conticelli and A. Peccerillo, "Petrology and Geochemistry of Potassic and Ultrapotassic Volcanism in Central Italy: Petrogenesis and Inferences on the Evolution of the Mantle Sources," Lithos, Vol. 28, No. 3-6, 1992, pp. 221- 240. doi:10.1016/0024-4937(92)90008-M

14. G. Marinelli, "Magma Evolution in Italy," In: C. H. Squyres, Ed., Geology of Italy, The Earth Science Society of the Lybian Arab Republic, Tripoli, 1975, pp. 165-219.

15. A. Peccerillo and P. Manetti, "The Potassium Alkaline Volcanism of Central-Southern Italy: A Review of the Data Relevant to Petrogenesis and Geodynamic Significance," Transactions of the Geological Society of South Africa, Vol. 88, No. 2, 1985, pp. 379-384.

16. C. Cimarelli and D. de Rita, "Structural Evolution of the Pleistocene Cimini Trachytic Volcanic Complex (Central Italy)," Bulletin of Volcanology, Vol. 68, No. 6, 2006, pp. 538-548. doi:10.1007/s00445-005-0028-3

17. D. Lardini and G. Nappi, "I Cicli Eruttivi del Complesso Vulcanico Cimino," Rendiconti della Società Italiana di Mineralogia e Petrologia, Vol. 42, 1987, pp. 141-153.

18. F. Sollevanti, "Geologic, Volcanologic and Tectonic Setting of the Vico-Cimino Area, Italy," Journal of Volcanology and Geothermal Research, Vol. 17, No. 1-4, 1983, pp. 203-217. doi:10.1016/0377-0273(83)90068-9

19. A. Bertagnini and A. Sbrana, "Il Vulcano di Vico: Stratigrafia del Complesso Vulcanico e Sequenze Eruttive Delle Formazioni Piroclastiche," Memorie della Società Geolologica Italiana, Vol. 35, No. 2, 1986, pp. 699-713.

20. G. Nappi, L. Valentini and M. Mattioli, "Ignimbritic Deposits in Central Italy: Pyroclastic Products of the Quaternary Age and Etruscan Footpaths. Field Trip Guide Book—P09," Proceedings 32nd International Geological Congress, Florence, 20-28 August 2004, p. 32.

21. G. Perini, S. Conticelli and L. Francalanci, "Inferences of the Volcanic History of the Vico Volcano, Roman Magmatic Province, Central Italy: Stratigraphic, Petrographic and Geochemical Data," Mineralogica et Petrographica Acta, Vol. 15, 1997, pp. 67-93.

22. G. Buonasorte, M. G. Carboni and M. A. Conti, "Il Substrato Plio-Pleistocenico Delle Vulcaniti Sabatine: Considerazioni Stratigrafiche e Paleoambientali," Bollettino della Società Geologica Italiana, Vol. 110, No. 1, 1991, pp. 35-40.

23. R. Funiciello, E. Locardi, G. Lombardi and M. Parotto, "The Main Volcanic Groups of Latium. Relations between Structural Evolution and Petrogenesis," Geologica Romana, Vol. 15, 1976, pp. 279-300.

24. P. La Torre, R. Nannini and F. Sollevanti, "Geothermal Exploration in Central Italy: Geophysical Survey in Cimini Range Area," European Association of Exploration Geophysicists: 43th Meeting, Venezia, 26-29 May 1981.

25. G. Buonasorte, A. Fiordelisi, E. Pandeli, U. Rossi and F. Sollevanti, "Stratigraphic Correlations and Structural Setting of the Pre-

Neoautochtonous Sedimentary Sequences of Northern Latium," Periodico di Mineralogia, Vol. 56, 1987, pp. 111-122.

26. A. Calamai, R. Cataldi, E. Locardi and A. Praturlon, "Distribuzione Delle Anomalie Geotermiche Nella Fascia Preappenninica Tosco-Laziale (Italia)," Symposium International Sobre Energia Geotérmica en America Latina, Ciudad de Guatemala, 16-23 October 1976, pp. 189-229.

27. B. Della Vedova, G. Pellis, J. P. Foucher and J. P Rehault, "Geothermal Structure of Tyrrhenian Sea," Marine Geology, Vol. 55, No. 3-4, 1984, pp. 271-289.doi:10.1016/0025-3227(84)90072-0

28. F. Mongelli, G. Zito, N. Ciaranfi and P. Pieri, "Interpretation of Heat Flow Density of the Apennine Chain, Italy," Tectonophysics, Vol. 164, No. 2-4, 1989, pp. 267-280.doi:10.1016/0040-1951(89)90020-6

29. G. Chiodini, F. Frondini and L. Marini, "Theoretical Geothermometers and P_{CO_2} Indicators for Aqueous Solutions Coming from Hydrothermal Systems of Medium-Low Temperature Hosted in Carbonate-Evaporite Rocks. Application to the Thermal Springs of the Etruscan Swell, Italy," Applied Geochemistry, Vol. 10, No. 3, 1995, pp. 337-346.doi:10.1016/0883-2927(95)00006-6

30. G. Chiodini, F. Frondini and F. Ponziani, "Deep Structures and Carbon Dioxide Degassing in Central Italy," Geothermics, Vol. 24, No. 1, 1995, pp. 81-94. doi:10.1016/0375-6505(94)00023-6

31. V. Duchi and A. Minissale, "Distribuzione Delle Manifestazioni Gassose Nel Settore Peritirrenico Tosco-Laziale e Loro Interazione con gli Acquiferi Superficiali," Bollettino della Società Geologica Italiana, Vol. 114, No. 2, 1995, pp. 337-351.

32. A. Minissale and V. Duchi, "Geothermometry on Fluids Circulating in a Carbonate Reservoir in North-Central Italy," Journal of Volcanology and Geothermal Research, Vol. 35, No. 3, 1988, pp. 237-252. doi:10.1016/0377-0273(88)90020-0

33. A. Minissale, D. M. Kerrick, G. Magro, M. T. Murell, M. Paladini, S. Rihs, N. C. Sturchio, F. Tassi and O. Vaselli, "Geochemistry of Quaternary Travertines in the Region North of Rome (Italy): Structural, Hydrologic and Paleoclimatic Implications," Earth

and Planetary Science Letters, Vol. 203, No. 2, 2002, pp. 709-728. doi:10.1016/S0012-821X(02)00875-0

34. A. Baiocchi, W. Dragoni, F. Lotti, G. Luzzi and V. Piscopo, "Outline of Theology of the Cimino and Vico Area and of the Interaction between Groundwater and Lake Vico (Lazio Region, Central Italy)," Bollettino della Società Geologica Italiana, Vol. 125, No. 2, 2006, pp. 187- 202.

35. C. Boni, P. Bono and G. Capelli, "Schema Idrogeologico Dell'Italia Centrale," Memorie della Società Geologica Italiana, Vol. 35, No. 2, 1986, pp. 991-1012.

36. G. Capelli, R. Mazza and C. Gazzetti, "Strumenti e Strategie per la Tutela e l'uso Compatibile Della Risorsa Idrica del Lazio: Gli Acquiferi Vulcanici," Pitagora, Bologna, 2005.

37. V. Piscopo, M. Barbieri, V. Monetti, G. Pagano, S. Pistoni, E. Ruggi and D. Stanzione, "Hydrogeology of Thermal Waters in Viterbo Area, Central Italy," Hydrogeology Journal, Vol. 14, No. 8, 2006, pp. 1508-1521. doi:10.1007/s10040-006-0090-8

38. U. Chiocchini, F. Castaldi, M. Barbieri and V. Eulilli, "A Stratigraphic and Geophysical Approach to Studying the Deep-Circulating Groundwater and Thermal Springs, and Their Recharge Areas, in Cimini Mountains-Viterbo Area, Central Italy," Hydrogeology Journal, Vol. 18, No. 6, 2010, pp. 1319-1341. doi:10.1007/s10040-010-0601-5

39. P. Baldi, G. C. Ferrara, L. Masselli and G. Pieretti, "Hydrogeochemistry of the Region between Monte Amiata and Rome," Geothermics, Vol. 2, No. 3-4, 1973, pp. 124- 141. doi:10.1016/0375-6505(73)90020-5

40. M. Angelone, C. Cremisini, V. Piscopo, M. Proposito and F. Spaziani, "Influence of Hydrostratigraphy and Structural Setting on the Arsenic Occurrence in Groundwater of the Cimino-Vico Volcanic Area (Central Italy)," Hydrogeology Journal, Vol. 17, No. 4, 2009, pp. 901-914. doi:10.1007/s10040-008-0401-3

41. G. Capasso and S. Inguaggiato, "A Simple Method for the Determination of Dissolved Gases in Natural Waters. An Application to Thermal Waters from Volcano Island," Applied Geochemistry, Vol. 13, No. 5, 1998, pp. 631-642. doi:10.1016/S0883-2927(97)00109-1

42.   S. Epstein and T. Mayeda, "Variation of ^{18}O Content of Water from Natural Sources," Geochimica et Cosmochimica Acta, Vol. 4, No. 5, 1953, pp. 213-224. doi:10.1016/0016-7037(53)90051-9

43.   F. Yanagisawa and H. Sakai, "Preparation of SO_2 for Sulfur Isotope Ratio Measurements by the Thermal Decomposition of $BaSO_4$-V_2O_5-SiO_2 Mixtures," Analytical Chemistry, Vol. 55, No. 6, 1983, pp. 985-987. doi:10.1021/ac00257a046

44.   M. P. Neary, "Tritium Enrichment—To Enrich or Not to Enrich?" Radioactivity and Radiochemistry, Vol. 8, 1997, pp. 23-35.

45.   SIMN, "Annali Idrologici," Servizio Idrografico e Mareografico Nazionale, Ministero dei Lavori Pubblici, Roma, 1951-1999.

46.   Regione Lazio, "Annali Idrologici," Ufficio Idrografico e Mereografico Regione Lazio, Roma, 2000-2010.

47.   G. Allen, S. Pereira, D. Raes and M. Smith, "Crop Evapotranspiration—Guidelines for Computing Crop Water Requirements," Natural Resources Management and Environment Department, FAO, Rome, 1998.

48.   C. T. Haan, "Statistical Methods in Hydrology," The Iowa State University Press, Ames, 1977.

49.   S. Carnicelli, D. Sagri, U. Chiocchini and S. Madonna, "Geopedologia," In: U. Chiocchini, Ed., La Geologia Della Città di Viterbo, Cangemi Editore, Roma, 2004.

50.   P. Lorenzoni, M. Raglione, P. Quantin, D. Bidini and L. Lulli, "Studio Dell'Apparato Vulcanico di Vico (Lazio). IV. I Suoli Delle Colate Piroclastiche," Annali Istituto Sperimentale di Studio e Difesa Suolo Firenze, Vol. 16, 1985, pp. 199-226.

51.   L. Lulli, C. Blasi, G. Abate, D. Bidini, S. Fascetti, P. Lorenzoni and M. Marchetti, "Studio Pedologico Dell'Apparato Vulcanico di Vico (Lazio). VIII. L'effetto Della Vegetazione Sulla Genesi dei Suoli," Annali Istituto Sperimentale di Studio e Difesa Suolo Firenze, Vol. 17, 1986, pp. 159-172.

52.   Regione Lazio, "Carta di Uso del Suolo," Direzione Regionale Territorio e Urbanistica, Regione Lazio, Roma, 2003.

53.   B. Conforto, "Risultati Della Prima Fase di Ricerche di Forze Endogene nel Viterbese," L'Ingegnere, Vol. 27, 1954, pp. 345-350.

54. B. Conforto, "Risultati Della Prima Fase di Ricerche di Forze Endogene nel Viterbese," L'Ingegnere, Vol. 27, 1954, pp. 521-530.

55. P. P. Mattias and V. Ventriglia, "La Regione Vulcanica dei Monti Cimini e Sabatini," Memorie della Società Geolologica Italiana, Vol. 9, No. 3, 1970, pp. 331-384.

56. B. Toro, "Anomalie Residue di Gravità e Strutture Profonde Nelle Aree Vulcaniche del Lazio Settentrionale," Geologica Romana, Vol. 17, 1978, pp. 35-44.

57. B. Camponeschi and F. Nolasco, "Le Risorse Naturali Della Regione Lazio. 2 Monti Cimini e Tuscia Romana," Regione Lazio, Roma, 1984.

58. G. Buonasorte, R. Cataldi, A. Ceccarelli, A. Costantini, S. D'Offizi, A. Lazzarotto, A., Ridolfi, P. Baldi, A. Borelli, G. Bertini, R. Bertrami, A. Calamai, G. Cameli, R. Corsi, C. D'Acquino, A. Fiordelisi, A. Grezzo and F. Lovari, "Ricerca ed Esplorazione Nell'Area Geotermica di Torre Alfina (Lazio—Umbria)," Bollettino della Società Geologica Italiana, Vol. 107, No. 2, 1988, pp. 265-337.

59. G. Buonasorte, E. Pandeli and A. Fiordelisi, "The Alfina 15 Well: Deep Geological Data from Northern Latium (Torre Alfina Geothermal Area)," Bollettino della Società Geologica Italiana, Vol. 110, No. 3-4, 1991, pp. 823-831.

60. R. Cataldi and M. Rendina, "Recent Discovery of a New Geothermal Field: Alfina," Geothermics, Vol. 2, No. 3-4, 1973, pp. 106-116. doi:10.1016/0375-6505(73)90016-3

61. G. Cavarretta, G. Giannelli, G. Scandiffio and F. Tecce, "Evolution of the Latera Geothermal System II: Metamorphic Hydrothermal Mineral Assemblages and Fluid Chemistry," Journal of Volcanology and Geothermal Research, Vol. 26, No. 3-4, 1985, pp. 337-364. doi:10.1016/0377-0273(85)90063-0

62. V. Duchi, L. Matassoni, F. Tassi and B. Nisi, "Studio Geochimico dei Fluidi (Acque e Gas) Circolanti Nella Regione Vulcanica dei M.ti Vulsini (Italia Centrale)," Bollettino della Società Geologica Italiana, Vol. 122, No. 1, 2003, pp. 47-61.

63. G. Giannelli and G. Scandiffio, "The Latera Geothermal System (Italy): Chemical Composition of the Geothermal Fluid and

Hypotheses on Its Origin," Geothermics, Vol. 18, No. 3, 1989, pp. 447-463. doi:10.1016/0375-6505(89)90068-0

64. G. F. Pinder and J. F. Jones, "Determination of the Groundwater Component of Peak Discharge from Chemistry of Total Runoff," Water Resources Research, Vol. 5, No. 2, 1969, pp. 438-445. doi:10.1029/WR005i002p00438

65. H. Craig, "Isotopic Variations in Meteoric Water," Science, Vol. 133, No. 3465, 1961, pp. 1702-1703. doi:10.1126/science.133.3465.1702

66. A. Longinelli and E. Selmo, "Isotopic Composition of Precipitation in Italy: A First Overall Map," Journal of Hydrology, Vol. 270, No. 1-2, 2003, pp. 75-88.

67. I. D. Clark and P. Fritz, "Environmental isotopes in hydrogeology," Lewis Publishers, New York, 1997.

68. C. W. Thornthwaite, "An Approach toward a Rational Classification of Climate," Geographical Review, Vol. 38, No. 1, 1948, pp. 55-94. doi:10.2307/210739

69. C. W. Thornthwaite and J. R. Mather, "The Water Balance," Publication in Climatology, Vol. 8, No. 1, 1955.

Chapter 8

Monitoring Recreational Waters: How to Integrate Environmental Determinants

Patricia Turgeon[1, 2]

[1]Laboratory for Foodborne Zoonoses, Public Health Agency of Canada, Saint-Hyacinthe, Canada

[2]Groupe de Recherche en Épidémiologie des Zoonoses et Santé Publique, Université de Montréal, Saint-Hyacinthe, Canada

ABSTRACT

Recreational waters are associated with a higher risk of disease for people engaged in activities that bring them into contact with these waters. The primary cause of contamination of recreational waters is fecal microorganisms, which may originate from various sources and involve several modulating factors, making it a complex public health and environmental issue. Monitoring recreational water quality should include two key components: Microbial water testing and monitoring environmental determinants associated with higher risks of contamination. Conducting both activities provides the foundation for a comprehensive assessment according to risk and the actual level of fecal pollution and thus could promote good management actions to ensure safe water quality. Nevertheless, monitoring of environmental determinants is rarely fully integrated in monitoring programs and is also harder to achieve, especially when water pollution is mainly associated with nonpoint sources. In order to achieve identification and monitoring of environmental determinants associated with fecal contamination of recreational waters, some specific steps should be followed and some questions must be answered. The objective of this review article is to present current knowledge on this topic and to suggest and discuss recommendations. Potential sources of contamination and factors able to modulate them should be identified and measured after the geographical area influencing fecal contamination of recreational water has been delineated. Statistical models have been developed to identify the relative importance of these environmental characteristics on fecal pollution of recreational waters but they do not allow for a full comprehension of the exact processes leading to this pollution, thus other methods should also be used to better understand these processes.

INTRODUCTION

In the last few years, health risks associated with activeties in recreational waters were a growing preoccupation for public health communities all around the world, but more specifically for industrialized countries. This interest can be explained in part by the growing attendance on public beaches, driven by climatic and demographic changes. In the next decades, global changes, including climate changes, will cause important perturbations to aquatic ecosystems. Among these, degradation of microbiological quality of surface waters is expected [1]. Climate changes will bring an increase in the frequency and intensity of rain events which will be followed by an increasing amount of water reaching watercourses by runoff [2]. This increased runoff will carry a larger amount of microorganisms. Moreover, the growing frequency of rain events could overload the sewer systems and the wastewater treatment plants, which were not designed to work with and treat this amount of water. These overloads could prevent the appropriate performance of treatment plants and thus lead to the dumping of wastewater directly into watercourses.

Human populations can be exposed to impaired surface water through drinking water but also through recreational waters. Different types of pollution can affect recreational waters, such as fecal contamination, cyanobacteria, and chemical pollutants. Fecal contamination can result in a large amount of microorganisms that lead to various types of illnesses including gastrointestinal illnesses (GI) and respiratory, skin, and ear infections [3]. GI illnesses are the most frequent diseases associated with activities in recreational waters but they are usually mild and self-limiting, which can lead to some difficulties in their monitoring. Nonetheless, epidemiological studies have shown positive associations between sporadic cases of GI illness and activities in recreational waters and between fecal contamination of these waters and GI outbreaks [4-8].

Surveillance of the microbial quality of recreational waters is mainly done by measuring levels of fecal microbial indicators such as Escherichia coli or fecal coliforms. This measurement can provide an indication of water quality relatively quickly, usually in 24 - 48 hours following the sampling. Different microbiological technologies like QPCR have been developed to palliate this time limit, providing results in only a few hours [9,10]. Moreover, predictive models have been elaborated from various meteorological parameters such as rainfall and wind speed in an attempt to predict the level of fecal contamination according to these conditions [11,12]. In general, these methods aim to answer the question: Are recreational waters safe for people today or in the next few days? Although these measures are crucial for quick decision making on the likelihood of a microbial hazard, they do not allow a complete evaluation of the risk of fecal contamination.

To account for a broader assessment of risk, the World Health Organization (WHO) recommends the evaluation and monitoring of sources of fecal contamination and environmental characteristics (environmental determinants) influencing this contamination in addition to water testing [13]. Unlike rainfall events and temperature, these characteristics are relatively stable in time and could explain as much as 40% of the fecal contamination variation for a beach [14].

Knowing and monitoring these determinants could help ascertain which beaches have a higher risk of fecal contamination. Integrating a sound assessment of both risk and determinants into the monitoring of fecal pollution using microbial indicators would provide the base for a more comprehensive and accurate evaluation and thus would promote good management actions to ensure safe water quality.

Monitoring of the sources of fecal contamination can be performed in different ways. The WHO recommends the annual census of every potential source of contamination of recreational waters by field inspection activities [13]. This precise and relevant method for point source pollution can be very difficult to apply to water bodies associated mostly with non-point source (diffuse) pollution, like most fresh and inland water bodies. Moreover, annual field campaigns can be time consuming and also very demanding in terms of financial and human resources, resulting in a big challenge when trying to apply them across large territories. Therefore, there is a need to develop new efficient methods able to characterize the proximal environment of beaches and thus contribute to the identification and monitoring of environmental characteristics associated with a higher risk of fecal contamination.

In order to achieve this identification and monitoring for recreational waters, some specific steps should be followed and some questions must be answered. The objective of this article is to conduct a review of knowledge on this topic and to suggest and discuss four steps that should be followed:

- Identify potential sources of fecal contamination of recreational waters.
- Identify factors able to modulate this pollution.
- Delineate geographical area influencing the fecal pollution of water bodies.
- Identify and apply methods for quantitatively measuring these sources and factors and assessing their association with fecal pollution of recreational waters in order to establish best monitoring practices.

IDENTIFY POTENTIAL SOURCES OF FECAL CONTAMINATION OF RECREATIONAL WATERS

Sources of fecal contamination are various and can include urban, wildlife, and agricultural activities. Water contaminated by both animal and human sources of fecal microorganisms can represent a health risk for people engaged in recreational activities. Although it is assumed that, in general, animal sources of fecal contamination represent a smaller risk to human health than human sources [13], animals can carry zoonotic pathogens such as Salmonella, Campylobacter, and E. coli O157:H7, which may lead to gastrointestinal illness and sometimes severe sequelae [15,16]. Studies have indicated that risk of gastrointestinal illness associated with exposure to recreational waters impacted by animal feces may not be different from waters impacted by human sources. Nonetheless, human health risk from exposure to recreational waters impacted by non-human sources is still not well understood; some studies did not find statistically significant associations between illness risk and fecal contamination of water by animal sources [17,18].

Urban Activities

Urban fecal pollution can occur through different means. During rain events, waters can runoff by two major ways. They can reach watercourses via pluvial sewer systems, carrying many pollutants such as organic components and fecal pathogens coming from domestic wastes, domestic fauna, and wildlife [19]. Waters can also be directed to combined sewer systems that collect domestic industrial and pluvial waters [20]. However, during heavy rainfalls and rapid snow melts, water discharge coming into these systems can overwhelm wastewater treatment plant capacities and overload. Water containing pluvial waters and wastewater from domestic and industrial use can be discharged into watercourses by overload drains called combined sewer overflows (CSOs) [21]. CSOs may represent a health risk for people in contact with these waters since the principal source of bacteria it contains is from human waste, which can contain large quantities of fecal pathogens [22]. E. coli concentration associated with CSOs can reach more than $10^6/100ml$, which it is much higher than recreational water guidelines [23]. Wastewater treated by wastewater treatment plants can also contain large amounts of fecal bacteria if these waters are not disinfected before being discharged into watercourses, placing people swimming in waters in close proximity to these discharges at higher risk for health problems [24].

Fecal pollution coming from urban activities and populations can also occur directly on the beaches. Bathers can be a source of fecal microorganisms via fecal accident, mainly where the proportion of children among bathers is high and if there are babies and toddlers in diapers [25,26]. Some outbreaks of Shigella sonnei (human pathogen) associated with activities in recreational freshwaters have been reported. In these outbreaks, the bacteria source mostly implicated were bathers who had been attending the beaches [27-29]. Bathers can also carry fecal microorganisms from the sand to the water and also stir up bottom sediment, making contact between bathers and the microorganisms trapped in sediment more frequent [30].

Domestic fauna such as dogs and cats can also be a source of fecal contamination. These animals can carry pathogens like species of Campylobacter, Salmonella, Giardia, and Cryptosporidium [31,32]. These microorganisms could reach watercourses mainly by runoff following rain events.

Wildlife

Wild animals can contribute to fecal pollution of recreational waters. Waterfowl have specifically been studied in the last decade [33,34]. These birds can be carriers of many fecal pathogens including species of Campylobacter, Salmonella, and Cryptosporidium and studies suggest they can be a source of contamination of recreational waters, particularly in rural regions [35-38]. Birds can pollute water by direct deposit of fecal material but also by contaminating the beach sand. Wild mammals can also carry fecal pathogens communicable to humans, although prevalence is usually low [39-41]. These animals can bring fecal microorganisms into watercourses by soil leaching and runoff during rain events and by direct deposit of fecal material.

Agricultural Activities

Agricultural activities can contribute to fecal pollution of waters and eventually recreational waters in many ways. Fecal pollution can come from animals and manure piles on farms and animal production sites. Livestock can carry and excrete in their feces zoonotic agents which may lead to various health problems from self-limiting GI disturbances to severe diseases that are potentially deadly. Zoonotic pathogens most often associated with farm animals are Salmonella Enterica, Campylobacter jejuni and coli, E. coli O157:H7, Giardia spp. and Cryptosporidium spp. [42,43]. These pathogens can reach watercourses mainly following rain events. Water contamination by stored manure on farms can result from leaking storage, overflows, or manure piles. A significant GI outbreak caused by E. coli O157:H7 and Campylobacter occurred in Walkerton, Canada in 2000 following a contamination of a municipal water well. Much of the evidence suggests that the origin of the contamination was runoff coming from a manure pile close to a cattle farm [44]. Manure spread on crop lands can also contain a large amount of fecal microorganisms [45]. As an example, fecal coliform concentrations measured in runoff waters following cattle manure spreading can reach 1.9×10^4 to 1.1×10^6/100ml, depending on the initial manure concentration [46]. Therefore, runoff waters coming from agricultural fields often exceed the drinking and microbiological guidelines for recreational waters [47]. Even if they can survive for many weeks on the soil after manure spreading, risk of water pollution by fecal microorganisms is highest soon after the spreading. On the other hand, some procedures can reduce the concentration of microorganisms in manure before spreading, such as aerobic or anaerobic digestion and composting [48,49].

In addition to manure spreading, animals at pasture can be responsible for water contamination through microorganisms present in their feces [50]. For instance, some agents can persist in soil for many months after exiting an animal's gastrointestinal tract [51,52]. The impact of animals at pasture can vary according to animal density. A high density can lead to a large quantity of urine and feces on a relatively small surface, increasing the probability of nutrients and microorganisms reaching surface waters through runoff and also getting through the soil to reach the ground waters [53]. High animal density can also accelerate soil erosion which can increase runoff during rain events. The impact of animals at pasture could also be associated with their distribution across the watershed and if they have access to a watercourse [54].

IDENTIFY FACTORS ABLE TO MODULATE FECAL POLLUTION OF RECREATIONAL WATERS

Some factors can modulate fecal contamination of waters by influencing the passage of fecal microorganisms to surface waters, including vegetation, climatic conditions, soil type, and topography.

Vegetation

Riparian zones can reduce the concentration of fecal microbes coming from agricultural lands and entering surface waters by more than 90% by preventing soil erosion and contributing to absorption of runoff water [55-57]. However, no scientific consensus exists on the minimal required width for an efficient reduction of microorganism transfer, although it is known that recommended width is affected by ground slope and vegetation type [58]. Wetlands can capture many elements brought by runoff such as sediments, microorganisms, and nutrients, reducing water pollution [59,60]. According to a study carried out in an agricultural watershed, wetlands could capture as much as 68% of E. coli coming from runoff waters [60]. Forest areas can also be associated with better water quality given their capacity to reduce runoff, sediments, and nutrients in surface waters. They could also have a positive influence on surface water quality by promoting water infiltration to the water tables [61].

Climatic Conditions

Rain events and their intensity are an important factor affecting the transport of microorganisms at the soil level. Runoff can carry bacteria a long distance downstream and thus contribute to water pollution [62-64]. As an example, risk of fecal contamination of watercourses following manure spreading during dry periods would be significantly lower than when manure is spread within a few days before rain events [64]. If there is no rain during days following manure spreading, bacteria movement to watercourses will diminish and under some environmental conditions—UV rays and desiccation—bacteria count will also diminish [65]. Moreover, during rainfall some bacteria such as E. coli would be quickly carried away by runoff waters and would have less chance to interact with the soil matrix [66]. Furthermore, studies have shown positives associations between rain events and waterborne disease outbreaks [67-69].

Exposure to solar rays could also be an important element in the bacterial inactivation process in water and soil [62]. Bacteria count in water decreases more rapidly during sunny days than during cloudy days, even at various depths [70]. In general, microorganism survival is prolonged in cold temperatures [71].

Soil Type and Topography

The two major modes of microorganism movement with water in soil are the infiltration to ground waters and runoff to surface waters. Relative proportions of these two modes depend on various elements including the type of soil [72,73]. Soil composition such as clay and soils with high levels of organic matter can influence the movement and survival of microorganisms. High clay concentrations can influence this survival by offering a protection against environmental stresses such as UV and desiccation. For example, E. coli counts in soil could decrease more rapidly in sandy soil than in clay soil [74]. Humidity level can also influence the microorganism movement in soil. The higher the soil humidity, the more contaminated the runoff will be following a rain event because the high humidity facilitates microorganism transport to watercourses [63].

Land slope also plays a role in microorganism movement. The more steep the slope, the more runoff will occur and at faster speeds, which promotes soil erosion and particle movement, which includes microorganisms [62,73].

DELINEATE GEOGRAPHICAL AREA INFLUENCING THE FECAL POLLUTION OF WATER BODIES

There are two ways by which surface waters and eventually recreational waters can be fecally polluted: they can be contaminated by point sources such as wastewater or storm water discharges; or they can be contaminated by non-point or diffuse sources. These sources can include some land uses as mentioned above—agricultural and urban lands. Diffuse pollution is harder to estimate and manage since it can be influenced by various characteristics of the drainage area such as topography and soil type and also by interactions between these characteristics and land uses.

Therefore when it is the time to identify and quantitatively evaluate the environmental characteristics or determinants that influence water quality, questions about the area influencing it the most remain not fully unanswered [75,76]. Those answers will inevitably vary according to the type of environment and type of water and will be different if we study surface waters in small vs. large watersheds, lakes vs. rivers, fresh vs. marine waters, or chemical vs. microbiological pollution. Despite the fact that a single answer is impossible to reach, since each water body and watershed has its own and unique combination of characteristics, few studies have tried to find definitive answers applicable to particular situations in microbiological pollution [14,76,77].

The study of Sliva et al. [76] compared two approaches to studying the impact of land use on chemical and microbiological water quality in three watersheds. Water sampling stations were located on rivers and microbiological quality was evaluated according to the fecal coliform level. The first approach was to measure landscape characteristics in an entire catchment area and to determine if there was a correlation with water quality measurements in the corresponding catchment. The second approach was to measure those same characteristics but only for a 100 m buffer around each sampling station. Results showed that the correlation between fecal coliforms and landscape predictors was different depending on the season of sampling. During the summer the correlation was better with the 100 m buffer zone data but during the fall, when the overland runoff is increased, the correlation was better with data on the entire catchment area.

Results of the study of Crowther et al. [77] also showed a difference in the relationship between land use and river water quality depending on if the sampling was done under baseor high-flow conditions. During baseflow conditions, E. coli levels seemed to be influenced more by land use within 2 km of the sampling station than the land within the whole catchment. Under highflow conditions, although the best correlation between E. coli levels and landscape data was seen at the whole catchment level, land use within 5 km was still able to explain more than 70% of the contamination variation. The study of Turgeon et al. [14] focused on recreational freshwaters associated with a lake and sampling was performed during the bathing season (mid-June to end of August). According to the results shown in Crowther et al., two buffer zones were delineated within the catchment area of each lake, one 2 km in size and another 5 km from the sampling point (beach) (Figure 1).Various land use and geo-hydrological characteristics were measured for each zone and the relationship between these characteristics and the fecal coliform levels was assessed. According to their results, the level of fecal coliform was influenced by agricultural activities and the additional 3 km of the 5 km buffer zone did not bring any additional information on the risk of contamination, suggesting that the most influence on water quality is from within the first 2 km.

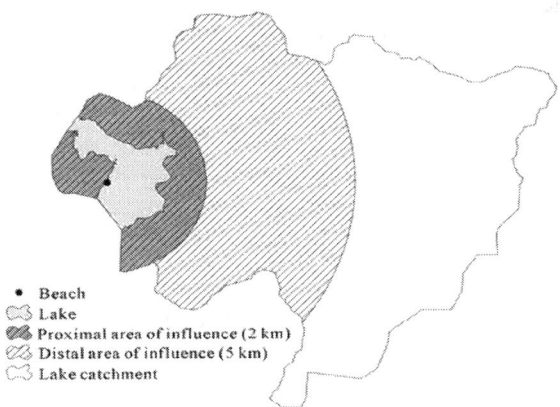

Figure 1: Two buffer zones used to delineate the geographical area influencing the fecal pollution of water bodies in the article of Turgeon et al. (With

kind permission from Springer Science +Business Media: Water Quality, Exposure and Health, Fecal Contamination of Recreational Freshwaters: the Effect of Time-Independent Factors, Vol. 3 (2), 2011, 109-118, P. Turgeon et al., Figure 1).

Although results of these studies differ according to their objectives and methods, one trend seems to emerge. During a dry season, microbiological surface water quality would be more influenced by land use characteristics located nearest the sampling station versus by the whole catchment, reflecting a possible die-off or sedimentation process along watercourses during this period.

QUANTITATIVELY MEASURING THESE SOURCES AND FACTORS AND THEIR ASSOCIATION WITH FECAL POLLUTION OF RECREATIONAL FRESHWATER

After having identified potential sources of fecal contamination and their modulating factors and delineating the area where they should be monitored more closely, the next step should be to quantitatively measure them in the areas of interest. Various sources of data can be used for this such as census data, regional or national thematic maps, and field survey data [14,78-80]. National census data from human populations can be useful for evaluating density or some characteristics such as age or socioeconomic distribution [81]. Census data from animal populations can also provide information on population density and species distribution or contribute to the evaluation of some management practices such as manure spreading [82]. The advantages of this data source are that data are usually standardized at the national level, available freely and for a long period of time. However, census data are collected most of the time by administrative regions or districts or centralized by farms or businesses.

These methods can bring spatial uncertainty which can be very relevant in the context of a hydrological process where the location of the source of contamination in regard to runoff direction and topography can play an important role [83,84]. Moreover, census data are usually constrained in regards to timeline. As an example, in Canada, population census is done every four years and agricultural census every five years, which can prevent availability of up-to-date data [81,82].

Data on landscape, land use, or on some environmental characteristics such as climatic conditions or topography can be acquired from regional or national thematic maps [80,85]. One advantage of using these maps is that one map is usually able to provide data on wide territories for more than one characteristic. Most of the time, data can be relatively easy to extract and thus can be ready to use with little manipulation. Nonetheless, like census processes, thematic maps are usually not performed every year and the timeline can be restrictive in a monitoring context. Furthermore, spatial resolution of these maps may limit their use or contribute to some measurement errors when data are needed for small areas [80]. Field surveys can provide accurate, precise, and up-to-date data on small areas, but one downside of this method is the high cost in terms of time and human and financial resources.

Lastly, land usage and landscape characteristics can also be provided by satellite imagery and more specifically Earth observation imagery [86,87]. These products can provide data on land use and land cover known to influence surface and recreational water quality such as agricultural lands, impervious (built) surfaces, wetlands, and forest areas [88-91]. This source of data has many advantages over more conventional sources. First, satellites can have a very large coverage and thus provide information on wide territories [92,93]. Satellite data from a particular sensor are always collected the same way and usually for a long period of time, allowing for a large amount of repeatable and consistent data which can be very relevant for a monitoring program [94]. Data from Earth observation can be highly precise depending on the sensor; knowing the precise data on characteristics associated with fecal contamination can be very relevant in the context of a hydrological matter. However, some factors can make their use difficult for recreational water monitoring, such as image cost and the expertise and technical resources needed for the gathering, processing, and analysing of those data [95-97]. As all sources of data present advantages and limitations, a combination of more than one source is probably the best way to tackle this step depending on the environment and the recreational waters involved.

Once the environmental characteristics associated with potential sources of fecal contamination and their modulator factors have been measured, it can be pertinent to try to identify which of these determinants have the most influence on recreational waters in the territory of interest. Most of the time, statistical methods that were used to achieve this step were multivariate regression modelling and Pearson's correlation coefficient [14,77-79,98]. Knowledge resulting from this step could provide the answer to the question: Which beaches have a higher risk of fecal contamination? Identification of those beaches combined with water sampling would provide the basis for an overall evaluation according to the risk and the actual level of fecal pollution, allowing for a better targeted monitoring of recreational waters.

CONCLUSION AND FUTURE DIRECTIONS

Based on a literature review, this paper aimed to suggest and discuss steps that should be followed for the integration of environmental determinants in the monitoring of recreational waters. Initially, potential sources of contamination and factors able to modulate them should be identified. Various sources and factors can be involved and depend on the region of interest. Measuring these sources and factors for monitoring purposes can only be done after delineating the geographical area influencing fecal contamination of recreational waters. No scientific consensus exists on this matter and few studies have participated in the debate. Nevertheless, in dry periods, which largely correspond to the bathing season, the proximal environment seems to play a more important role in the pollution process than the whole catchment does. Studies using mathematical and hydrological modelling would certainly help to bring new knowledge on this topic. Various sources of data can be used to provide information on sources of contamination and modulating factors and could be used to measure them on the geographical area of interest. Since none of them is sufficient in itself to provide all the information needed to fully characterize the proximal environment of recreational waters, a combination of more than one would be the best way to measure environmental characteristics. Some studies have used statistical models to identify the relative importance of environmental characteristics on fecal pollution of recreational waters. Although these models provide relevant information on environmental characteristics associated with a higher risk of fecal contamination and thus can serve as a good working basis for the integration of the monitoring of the environmental determinants of recreational water quality, they do not allow for a full comprehension of the exact processes leading to this pollution. Source attribution methods should be used to better understand these processes and to provide information on a better estimation of the precise contribution of each source of contamination.

Monitoring environmental determinants of water quality should be integrated with monitoring programs of recreational waters. In addition to water testing, it would provide an overall evaluation according to the risk of contamination and the actual level of fecal pollution, promoting in turn good water management actions to ensure safe water quality [13]. Moreover, knowing and monitoring these determinants could promote preventive actions to diminish their impact on water quality. In addition to contributing to better protecting the public, this preventive approach would align with the multi-barrier approach promoted in the domain of water quality [99, 100].

ACKNOWLEDGEMENTS

The author wishes to thank Dr. Pascal Michel of the Public Health Agency of Canada for his helpful comments on this paper.

REFERENCES

1. J. B. Rose, P. R. Epstein, E. K. Lipp, B. H. Sherman, S. M. Bernard and J. A. Patz, "Climate Variability and Change in the United States: Potential Impacts on Waterand Foodborne Diseases Caused by Microbiologic Agents [Review]," Environmental Health Perspectives, Vol. 109, Suppl. 2, 2001, pp. 211-221.

2. Intergovernmental Panel on Climate Change, "Climate Change 2007: Synthesis Report," Valencia, Spain, 2007. http://www.ipcc.ch/pdf/assessment-report/ar4/syr/ar4_syr.pdf

3. A. Pruss, "Review of Epidemiological Studies on Health Effects from Exposure to Recreational Water," International Journal

of Epidemiology, Vol. 27, No. 1, 1998, pp. 1-9.doi:10.1093/ije/27.1.1

4.   K. A. Feldman, J. C. Mohle-Boetani, J. Ward, K. Furst, S. L. Abbott, D. V. Ferrero, A. Olsen and S. B. Werner, "A Cluster of Escherichia coli O157: Nonmotile Infections Associated with Recreational Exposure to Lake Water," Public Health Reports, Vol. 117, No. 4, 2002, pp. 380- 385.

5.   M. G. Bruce, M. B. Curtis, M. M. Payne, R. K. Gautom, E. C. Thompson, A. L. Bennett and J. M. Kobayashi, "Lake-Associated Outbreak of Escherichia coli O157:H7 in Clark County, Washington, August 1999," Archives of Pediatrics & Adolescent Medicine, Vol. 157, No. 10, 2003, pp. 1016-1021. doi:10.1001/archpedi.157.10.1016

6.   A. Wiedenmann, P. Kruger, K. Dietz, J. M. Lopez-Pila, R. Szewzyk and K. Botzenhart, "A Randomized Controlled Trial Assessing Infectious Disease Risks from Bathing in Fresh Recreational Waters in Relation to the Concentration of Escherichia coli, Intestinal Enterococci, Clostridium perfringens and Somatic coliphages," Environmental Health Perspectives, Vol. 114, No. 2, 2006, pp. 228-236. doi:10.1289/ehp.8115

7.   B. Sartorius, Y. Andersson, I. Velicko, B. De Jong, M. Löfdahl, K.-O. Hedlund, G. Allestam, C. Wangsell, O. Bergstedt, P. Horal, P. Ulleryd and A. Soderstrom, "Outbreak of Norovirus in Västra Götaland Associated with Recreational Activities at Two Lakes during August 2004," Scandinavian Journal of Infectious Diseases, Vol. 39, No. 4, 2007, pp. 323-331. doi:10.1080/00365540601053006

8.   J. M. Fleisher, L. E. Fleming, H. M. Solo-Gabriele, J. K. Kish, C. D. Sinigalliano, L. Plano, S. M. Elmir, J. D. Wang, K. Withum, T. Shibata, M. L. Gidley, A. Abdelzaher, G. Q. He, C. Ortega, X. F. Zhu, M. Wright, J. Hollenbeck and L. C. Backer, "The BEACHES Study: Health Effects and Exposures from Non-Point Source Microbial Contaminants in Subtropical Recreational Marine Waters," International Journal of Epidemiology, Vol. 39, No. 5, 2010, pp. 1291-1298. doi:10.1093/ije/dyq084

9.   J. W. Santo Domingo, S. C. Siefring and R. A. Haugland, "Real-Time PCR Method to Detect Enterococcus faecalis in Water,"

Biotechnology Letters, Vol. 25, No. 3, 2003, pp. 261-265. doi:10.1023/A:1022303118122

10. T. J. Wade, R. L. Calderon, E. Sams, M. Beach, K. P. Brenner, A. H. Williams and A. P. Dufour, "Rapidly Measured Indicators of Recreational Water Quality Are Predictive of Swimming-Associated Gastrointestional Illness," Environmental Health Perspectives, Vol. 114, No. 1, 2006, pp. 24-28. doi:10.1289/ehp.8273

11. G. A. Olyphant and R. L. Whitman, "Elements of a Predictive Model for Determining Beach Closures on a Real Time Basis: The Case of 63rd Street Beach Chicago," Environmental Monitoring & Assessment, Vol. 98, No. 1-3, 2004, pp. 175-190.doi:10.1023/B:EMAS.0000038185.79137.b9

12. D. S. Francy, R. A. Darner and E. E. Bertke, "Models for Predicting Recreational Water Quality at Lake Erie Beaches," US Geological Survey. Scientific Investigations Report 2006-5192, 2006.

13. World Health Organization, "Guidelines for Safe Recreationnal Water Environments. Coastal and Fresh Waters," Geneva, Switzerland, 2003, http://www.who.int/water_sanitation_health/bathing/srwe1/en/

14. P. Turgeon, P. Michel, P. Levallois, M. Archambault and A. Ravel, "Fecal Contamination of Recreational Freshwaters: The Effect of Time-Independent Agroenvironmental Factors," Water Quality, Exposure and Health, Vol. 3, No. 2, 2011, pp. 109-118. doi:10.1007/s12403-011-0048-5

15. M. L. Hutchison, L. D. Walters, S. M. Avery, B. A. Synge and A. Moore, "Levels of Zoonotic Agents in British Livestock Manures," Letters in Applied Microbiology, Vol. 39, No. 2, 2004, pp. 207-214. doi:10.1111/j.1472-765X.2004.01564.x

16. E. M. Moriarty, L. W. Sinton, M. L. Mackenzie, N. Karki and D. R. Wood, "A Survey of Enteric Bacteria and Protozoans in Fresh Bovine Faeces on New Zealand Dairy Farms," Journal of Applied Microbiology, Vol. 105, No. 6, 2008, pp. 2015-2025. doi:10.1111/j.1365-2672.2008.03939.x

17. R. Calderon, M. E. and A. Dufour, "Health Effects of Swimmers and Non-Point Sources of Contaminated Water," International Journal of Environmental Health Research, Vol. 1, 1991, pp. 21-31.

18.   J. M. Colford Jr., T. J. Wade, K. C. Schiff, C. C. Wright and J. F. Griffith, "Water Quality Indicators and the Risk of Illness at Beaches with Non-Point Sources of Fecal Contamination," Epidemiology, Vol. 18, 2007, pp. 27-35. doi:10.1097/01.ede.0000249425.32990.b9

19.   V. P. Olivieri, K. Kawata and S. H. Lim, "Microbiological Impacts of Storm Sewer Overflows: Some Aspects of the Implication of Microbiological Indicators for Receiving Waters," In: J. B. Ellis, Ed., Urban Discharges and Receiving Water Quality Impacts, Pergamon Press, Oxford, 1989, pp. 47-54.

20.   Environment Canada, "Wastewater Management," Environment Canada,   2009.   http://www.ec.gc.ca/eu-ww/default. asp?lang=En&n=0FB32EFD-1

21.   United States Environmental Protection Agency, "Report to Congress. Impacts and Control of CSOs and SSOs," Washington DC, 2004. http://cfpub.epa.gov/npdes/cso/cpolicy_report2004. cfm

22.   J. A. Castro-Hermida, I. Garcia-Presedo, A. Almeida, M. Gonzalez-Warleta, J. M. C. Da Costa and M. Mezo, "Contribution of Treated Wastewater to the Contamination of Recreational River Areas with Cryptosporidium spp. and Giardia duodenalis," Water Research, Vol. 42, No. 13, 2008, pp. 3528-3538. doi:10.1016/j. watres.2008.05.001

23.   J. Marsalek and Q. Rochfort, "Urban Wet-Weather Flows: Sources of Fecal Contamination Impacting on Recreational Waters and Threatening Drinking-Water Sources," Journal of Toxicology and Environmental Health, Vol. 67, No. 20-22, 2004, pp. 1765-7717. doi:10.1080/15287390490492430

24.   P. Payment, R. Plante and P. Cejka, "Removal of Indicator Bacteria, Human Enteric Viruses, Giardia Cysts, and Cryptosporidium Oocysts at a Large Wastewater Primary Treatment Facility," Canadian Journal of Microbiology, Vol. 47, No. 3, 2001, pp. 188-193.

25.   C. P. Gerba, "Assessment of Enteric Pathogen Shedding by Bathers during Recreational Activity and its Impact on Water Quality," Quantitative Microbiology, Vol. 2, 2000, pp. 55-68.

26.   D. Sunderland, T. K. Graczyk, L. Tamang and P. N. Breysse, "Impact of Bathers on Levels of Cryptosporidium parvum Oocysts and Giardia lamblia Cysts in Recreational Beach Waters," Water

Research, Vol. 41, No. 15, 2007, pp. 3483-3489.doi:10.1016/j. watres.2007.05.009

27. J. Blostein, "Shigellosis from Swimming in a Park Pond in Michigan," Public Health Reports, Vol. 106, No. 3, 1991, pp. 317-322.

28. W. E. Keene, J. M. McAnulty, F. C. Hoesly, L. P. Williams, K. Hedberg, G. L. Oxman, T. J. Barrett, M. A. Pfaller and D. W. Fleming, "A Swimming-Associated Outbreak of Hemorrhagic Colitis Caused by Escherichia coli O157:H7 and Shigella Sonnei," New England Journal of Medicine, Vol. 331, No. 9, 1994, pp. 579-584.doi:10.1056/NEJM199409013310904

29. M. Iwamoto, G. Hlady, M. Jeter, C. Burnett, C. Drenzek, S. Lance, J. Benson, D. Page and P. Blake, "Shigellosis among Swimmers in a Freshwater Lake," Southern Medical Journal, Vol. 98, No. 8, 2005, pp. 774-778. doi:10.1097/01.smj.0000172764.14147.e5

30. T. K. Graczyk, D. Sunderland, G. N. Awantang, Y. Mashinski, F. E. Lucy, Z. Graczyk, L. Chomicz and P. N. Breysse, "Relationships among Bather Density, Levels of Human Waterborne Pathogens, and Fecal Coliform Counts in Marine Recreational Beach Water," Journal of Parasitology Research, Vol. 106, No. 5, 2010, pp. 1103- 1108.doi:10.1007/s00436-010-1769-2

31. S. Hill, J. M. Cheney, G. F. Taton-Allen, J. S. Reif, C. Bruns and M. R. Lappin, "Prevalence of Enteric Zoonotic Organisms in Cats," Journal of the American Vetrinary Medical Association, Vol. 216, No. 5, 2000, pp. 687-692. doi:10.2460/javma.2000.216.687

32. T. Hackett and M. R. Lappin, "Prevalence of Enteric Pathogens in Dogs of North-Central Colorado," Journal of the American Animal Hospital Association, Vol. 39, 2003, pp. 52-56.

33. L. R. Fogarty, S. K. Haack, M. J. Wolcott and R. L. Whitman, "Abundance and Characteristics of the Recreational Water Quality Indicator Bacteria Escherichia coli and Enterococci in Gull Faeces," Journal of Applied Microbiology, Vol. 94, No. 5, 2003, pp. 865-878. doi:10.1046/j.1365-2672.2003.01910.x

34. S. K. Haack, L. R. Fogarty and C. Wright, "Escherichia coli and Enterococci at Beaches in the Grand Traverse Bay, Lake Michigan: Sources, Characteristics, and Environmental Pathways," Environmental Science & Technology, Vol. 37, No. 15, 2003, pp. 3275-3282.doi:10.1021/es021062n

35. Z. Hubalek, "An Annotated Checklist of Pathogenic Microorganisms Associated with Migratory Birds [Review]," Journal of Wildlife Diseases, Vol. 40, No. 4, 2004, pp. 639-659.

36. K. J. Meyer, C. M. Appletoft, A. K. Schwemm, K. Uzoigwe and E. J. Brown, "Determining the Source of Fecal Contamination in Recreational Waters," Journal of Environmental Health, Vol. 68, No. 1, 2005, pp. 25-30.

37. T. A. Edge and S. Hill, "Multiple Lines of Evidence to Identify the Sources of Fecal Pollution at a Freshwater Beach in Hamilton Harbour, Lake Ontario," Water Research, Vol. 41, No. 16, 2007, pp. 3585-3594. doi:10.1016/j.watres.2007.05.012

38. T. K. Graczyk, A. C. Majewska and K. J. Schwab, "The Role of Birds in Dissemination of Human Waterborne Enteropathogens," Trends in Parasitology, Vol. 24, No. 2, 2008, pp. 55-59. doi:10.1016/j.pt.2007.10.007

39. V. R. Simpson, "Wild Animals as Reservoirs of Infectious Diseases in the UK," The Veterinary Journal, Vol. 163, No. 2, 2002, pp. 128-146. doi:10.1053/tvjl.2001.0662

40. K. Handeland, L. L. Nesse, A. Lillehaug, T. Voikoren, B. Djonne and B. Bergsjo, "Natural and Experimental Salmonella typhimurium Infections in Foxes (Vulpes vulpes)," Veterinary Microbiology, Vol. 132, No. 1-2, 2008, pp. 129-134. doi:10.1016/j.vetmic.2008.05.002

41. C. Jardine, R. J. Reid-Smith, N. Janecko, M. Allan and S. A. McEwen, "Salmonella in Racoons (Proycon lotor) in Southern Ontario, Canada," Journal of Wildlife Diseases, Vol. 47, No. 2, 2011, pp. 344-351.

42. R. Majdoub, C. Côté, M. Labadi, K. Guay and M. Généreux, "Impact de l'Utilisation des Engrais de Ferme sur la Qualité Microbiologique de l'eau Souterraine," Instituts de Recherche et de Développement en AgroEnvironnement, Québec, 2003.

43. P. Chevalier, P. Levallois and P. Michel, "Infections Entériques d'Origine Hydrique Potentiellement Associées à la Production Animale: Revue de la Littérature," Vecteur Environnement, Vol. 37, No. 2, 2004, pp. 90-106.

44. Government of Canada, "Waterborne Outbreak of Gastroenteritis Associated with a Contaminated Municipal Water Supply,

Walkerton, Ontario, May-June 2000," Canada Communicable Disease Report, Vol. 26, No. 20, 2000, pp. 170-173.

45. J. A. Thurston-Enriquez, J. E. Gilley and B. Eghball, "Microbial Quality of Runoff Following Land Application of Cattle Manure and Swine Slurry," Journal of Water & Health, Vol. 3, No. 2, 2005, pp. 157-171.

46. M. C. Ramos, J. N. Quinton and S. F. Tyrrel, "Effects of Cattle Manure on Erosion Rates and Runoff Water Pollution by Faecal Coliforms," Journal of Environmental Management, Vol. 78, No. 1, 2006, pp. 97-101. doi:10.1016/j.jenvman.2005.04.010

47. Health Canada, "Guidelines for Canadian Recreationnal Water Quality," Ottawa, 1992. http://www.hc-sc.gc.ca/ewh-semt/pubs/water-eau/guide_water-1992-guide_eau/index-eng.php

48. X. P. Jiang, J. Morgan and M. P. Doyle, "Fate of Escherichia coli O157:H7 during Composting of Bovine Manure in a Laboratory-Scale Bioreactor," Journal of Food Protection, Vol. 66, No. 1, 2003, pp. 25-30.

49. L. M. Avery, K. Killham and D. L. Jones, "Survival of E. coli O157:H7 in Organic Wastes Destined for Land Application," Journal of Applied Microbiology, Vol. 98, No. 4, 2005, pp. 814-22. doi:10.1111/j.1365-2672.2004.02524.x

50. A. J. A. Vinten, J. T. Douglas, D. R. Lewis, M. N. Aitken and D. R. Fenlon, "Relative Risk of Surface Water Pollution by E. coli Derived from Faeces of Grazing Animals Compared to Slurry Application," Soil Use & Management, Vol. 20, No. 1, 2004, pp. 13-22.doi:10.1079/SUM2004214

51. S. M. Avery, A. Moore and M. L. Hutchison, "Fate of Escherichia coli Originating from Livestock Faeces Deposited Directly onto Pasture," Letters in Applied Microbiology, Vol. 38, No. 5, 2004, pp. 355-359. doi:10.1111/j.1472-765X.2004.01501.x

52. L. W. Sinton, R. R. Braithwaite, C. H. Hall and M. L. Mackenzie, "Survival of Indicator and Pathogenic Bacteria in Bovine Feces on Pasture," Applied & Environmental Microbiology, Vol. 73, No. 24, 2007, pp. 7917- 7925. doi:10.1128/AEM.01620-07

53. R. K. Hubbard, G. L. Newton and G. M. Hill, "Water Quality and the Grazing Animals," Journal of animal science, Vol. 82, 2004, pp. E255-E263.

54. P. Rodgers, C. Soulsby, C. Hunter and J. Petry, "Spatial and Temporal Bacterial Quality of a Lowland Agricultural Stream in Northeast Scotland," The Science of the total Environment, Vol. 314-316, 2003, pp. 289-302. doi:10.1016/S0048-9697(03)00061-5

55. J. A. Entry, R. K. Hubbard, J. E. Thies and J. J. Fuhrmann, "The Influence of Vegetation in Riparian Filterstrips on Coliform Bacteria: I. Movement and Survival in Water," Journal of Environmental Quality, Vol. 29, No. 4, 2000, pp. 1206-1214. doi:10.2134/jeq2000.00472425002900040026x

56. R. M. Roodsari, D. R. Shelton, A. Shirmohammadi, Y. A. Pachepsky, A. M. Sadeghi and J. L. Starr, "Fecal Coliform Transport as Affected by Surface Condition," Transactions of the ASAE, Vol. 48, No. 3, 2005, pp. 1055-1061.

57. T. J. Sullivan, J. A. Moore, D. R. Thomas, E. Mallery, K. U. Snyder, M. Wustenberg, J. Wustenberg, S. D. Mackey and D. L. Moore, "Efficacy of Vegetated Buffers in Preventing Transport of Fecal Coliform Bacteria from Pasturelands," Environmental Management, Vol. 40, No. 6, 2007, pp. 958-965. doi:10.1007/s00267-007-9012-3

58. É. Gagnon and G. Gangbazo, "Efficacité des Bandes Riveraines: Analyse de la Documentation Scientifique et Perspectives," Ministère du Développment Durable, de l'Environnement et des Parcs du Québec, Québec, 2007.

59. C. Kao and M. Wu, "Control of Non-Point Source Pollution by a Natural Wetlands," Water Science and Technology, Vol. 43, No. 5, 2001, pp. 169-174.

60. [61] A. K. Knox, A. R. Dahlgren, K. W. Tate and E. R. Atwill, "Efficacy of Natural Wetlands to Retain Nutrient, Sediment and Microbial Polluants," Journal of Environmental Quality, Vol. 37, 2008, pp. 1837-1846. doi:10.2134/jeq2007.0067

61. [62] M. Matteo, T. Randhir and D. Bloniarz, "WatershedScale Impacts of Forest Buffers on Water Quality and Runoff in Urbanizing Environment," Journal of Water Resources Planning and Management, Vol. 132, No. 3, 2006, pp. 144-152. doi:10.1061/(ASCE)0733-9496(2006)132:3(144)

62. [63] J. Abu-Ashour and H. Lee, "Transport of Bacteria on Sloping Soil Surfaces by Runoff," Environmental Toxicology,

Vol. 15, No. 2, 2000, pp. 149-153.doi:10.1002/(SICI)1522-7278(2000)15:2<149::AID-TOX11>3.0.CO;2-O

63.   [64] C. Ferguson, A. M. D. Husman, N. Altavilla, D. Deere and N. Ashbolt, "Fate and Transport of Surface Water Pathogens in Watersheds [Review]," Critical Reviews in Environmental Science & Technology, Vol. 33, No. 3, 2003, pp. 299-361. doi:10.1080/10643380390814497

64.   [65] I. D. Ogden, D. R. Fenlon, A. J. A. Vinten and D. Lewis, "The fate of Escherichia coli O157 in Soil and Its Potential to Contaminate Drinking Water," International Journal of Food Microbiology, Vol. 66, No. 1-2, 2001, pp. 111-117. doi:10.1016/S0168-1605(00)00508-0

65.   [66] D. Trevisan, J. Y. Vansteelant and J. M. Dorioz, "Survival and Leaching of Fecal Bacteria after Slurry Spreading on Mountain Hay Meadows: Consequences for the Management of Water Contamination Risk," Water Research, Vol. 36, No. 1, 2002, pp. 275-283. doi:10.1016/S0043-1354(01)00184-1

66.   [67] R. W. Muirhead, R. P. Collins and P. J. Bremer, "Interaction of Escherichia coli and Soil Particles in Runoff," Applied and Environmental Microbiology, Vol. 72, No. 5, 2006, pp. 3406-3411. doi:10.1128/AEM.72.5.3406-3411.2006

67.   [68] F. C. Curriero, J. A. Patz, J. B. Rose and S. Lele, "The Association between Extreme Precipitation and Waterborne Disease Outbreaks in the United States, 1948- 1994," American Journal of Public Health, Vol. 91, No. 8, 2001, pp. 1194-1199. doi:10.2105/AJPH.91.8.1194

68.   [69] D. F. Charron, M. K. Thomas, D. Waltner-Toews, J. J. Aramini, T. Edge, R. A. Kent, A. R. Maarouf and J. Wilson, "Vulnerability of Waterborne Diseases to Climate Change in Canada: A Review," Journal of Toxicology & Environmental Health Part A, Vol. 67, No. 20-22, 2004, pp. 1667-1677. doi:10.1080/15287390490492313

69.   [70] J. A. Patz, S. J. Vavrus, C. K. Uejio and S. L. McLellan, "Climate Change and Waterborne Disease Risk in the Great Lakes Region of the US," American Journal of Preventive Medicine, Vol. 35, No. 5, 2008, pp. 451-458.doi:10.1016/j.amepre.2008.08.026

70.   [71] R. L. Whitman, M. B. Nevers, G. C. Korinek and M. N. Byappanahalli, "Solar and Temporal Effects on Escherichia coli Concentration at a Lake Michigan Swimming Beach," Applied

and Environmental Microbiology, Vol. 70, No. 7, 2004, pp. 4276-4285.doi:10.1128/AEM.70.7.4276-4285.2004

71. [72] S. R. Crane and J. A. Moore, "Modeling Enteric Bacterial Die-Off: A Review," Water Air soil Pollution, Vol. 27, 1986, p. 411. doi:10.1007/BF00649422

72. [73] J. Abu-Ashour, D. M. Joy, H. Lee, H. R. Whiteley and S. Zelin, "Transport of Microorganisms through Soil," Water, Air, & Soil Pollution, Vol. 75, 1994, pp. 141-158.doi:10.1007/BF01100406

73. [74] R. C. Jamieson, R. J. Gordon, K. E. Sharples, G. W. Stratton and A. Madani, "Movement and Persistence of Fecal Bacteria in Agricultural Soils and Subsurface Drainage Water: A Review," Canadian Biosystems Engineering, Vol. 44, No. 1, 2002, pp. 1-9.

74. [75] M. M. Lau, S. C. Ingham and A. R. Arment, "Survival of Faecal Indicator Bacteria in Bovine Manure Incorporated into Soil," Letters in Applied Microbiology, Vol. 33, 2001, p. 131.

75. [76] L. Johnson and S. Gage, "Landscape Approaches to the Analysis of Aquatic Ecosystems," Freshwater Biology, Vol. 37, No. 1, 1997, pp. 113-132.doi:10.1046/j.1365-2427.1997.00156.x

76. [77] L. Sliva and D. D. Williams, "BUFFER zone versus Whole Catchment Approaches to Studying Land Use Impact on River Water Quality," Water Research, Vol. 35, No. 14, 2001, pp. 3462-3472. doi:10.1016/S0043-1354(01)00062-8

77. [78] J. Crowther, M. D. Wyer, M. Bradford, D. Kay, C. A. Francis and W. G. Knisel, "Modelling Faecal Indicator Concentrations in Large Rural Catchments Using Land Use and Topographic Data," Journal of Applied Microbiology, Vol. 94, No. 6, 2003, pp. 962-973.doi:10.1046/j.1365-2672.2003.01877.x

78. [79] J. Crowther, D. Kay and M. D. Wyer, "Faecal-Indicator Concentrations in Waters Draining Lowland Pastoral Catchments in the UK: Relationships with Land Use and Farming Practices," Water Research, Vol. 36, No. 7, 2002, pp. 1725-1734. doi:10.1016/S0043-1354(01)00394-3

79. [80] D. Kay, M. Wyer, J. Crowther, C. Stapleton, M. Bradford, A. McDonald, J. Greaves, C. Francis and J. Watkins, "Predicting Faecal Indicator Fluxes Using Digital Land Use Data in the UK's Sentinel Water Framework Directive Catchment: The Ribble Study," Water Research, Vol. 39, No. 16, 2005, pp. 3967-3981. doi:10.1016/j.watres.2005.07.006

80. [81] D. Kay, S. Anthony, J. Crowther, B. J. Chambers, F. A. Nicholson, D. Chadwick, C. M. Stapleton and M. D. Wyer, "Microbial Water Pollution: A Screening Tool for Initial Catchment-Scale Assessment and Source Apportionment," Science of the Total Environment, Vol. 408, No. 23, 2010, pp. 5646-5656.doi:10.1016/j.scitotenv.2009.07.033

81. [82] Statistics Canada, "2006 Census," Statistics Canada, 2007. http://www12.statcan.gc.ca/census-recensement/index-eng.cfm

82. [83] Statistics Canada, "2006 Agriculture Census," Statistics Canada, 2008. http://www.statcan.gc.ca/ca-ra2006/index-eng.htm

83. [84] P. D'Arcy and R. Carignan, "Influence of Catchment Topography on Water Chemistry in Southeastern Quebec Shield Lakes," Canadian Journal of Fisheries & Aquatic Sciences, Vol. 54, 1997, pp. 2215-2227.

84. [85] P. Turgeon, P. Michel, P. Levallois, P. Chevalier, D. Daignault, B. Crago, R. Irwin, S. A. McEwen, N. F. Neumann and M. Louie, "Agroenvironmental Determinants Associated with the Presence of AntimicrobialResistant Escherichia coli in Beach Waters in Quebec, Canada," Zoonoses and Public Health, Vol. 58, No. 6, 2011, pp. 432-439.

85. [86] D. W. McKenney, M. F. Hutchinson, J. L. Kesteven and L. A. Venier, "Canada's Plant Hardiness Zones Revisited Using Modern Climate Interpolation Techniques," Canadian Journal of Plant Sciences, Vol. 81, 2001, pp. 129- 143.

86. [87] J. D. Boone, K. C. McGwire, E. W. Otteson, R. S. DeBaca, E. A. Kuhn, P. Villard, P. F. Brussard and S. C. St Jeor, "Remote Sensing and Geographic Information Systems: Charting Sin Nombre Virus Infections in Deer Mice," Emerging Infectious Diseases, Vol. 6, No. 3, 2000, pp. 248-258.

87. [88] A. Leblond, A. Sandoz, G. Lefebvre, H. Zeller and D. J. Bicout, "Remote Sensing Based Identification of Environmental Risk Factors Associated with West Nile Disease in Horses in Camargue, France," Preventive Veterinary Medicine, Vol. 79, No. 1, 2007, pp. 20-31.

88. [89] L. R. Beck, B. M. Lobitz and B. L. Wood, "Remote Sensing and Human Health: New Sensors and New Opportunities," Emerging Infectious Diseases, Vol. 6, No. 3, 2000, pp. 217-227.

89.   [90] S. Kalluri, P. Gilruth, D. Rogers and M. Szczur, "Surveillance of Arthropod Vector-Borne Infectious Diseases Using Remote Sensing Techniques: A Review Art. No. e116 [Review]," PLoS Pathogens, Vol. 3, No. 10, 2007, pp. 1361-1371.

90.   [91] J. B. Campbell, "Introduction to Remote Sensing," The Guilford Press, New York, 2007.

91.   [92] Y. Zhang, B. Guindon, K. Sun and L. Sun, "Remote Sensing for Improving Understanding on Canadian Urbanization," Canada Centre for Remote Sensing, Natural Resources Canada, 2010. http://www.ccrs.nrcan.gc.ca/optical/curlus_e.php

92.   [93] J. P. Messina and K. A. Crews-Meyer, "A Historical Perspective on the Development of Remotely Sensed Data as Applied to Medical Geography," In: D. P. Albert, W. M. Geslier and B. Levrgood, Eds., Spatial Analysis, GIS, and Remote Sensing Applications in the Health Sciences, Ann Arbor Press, Chelsea, 2000, pp. 129-146.

93.   [94] United Nations Office for Outer Space Affairs, "Space Solutions for the World's Problems," Vienna, 2005. http://www.oosa.unvienna.org/pdf/publications/IAM2005E.pdf

94.   [95] S. J. Goetz, S. D. Prince and J. Small, "Advances in Satellite Remote Sensing of Environmental Variables for Epidemiological Applications," In: S. I. Hay, S. E. Randolph and D. J. Rogers, Eds., Remote Sensing and Geographical Information Systems in Epidemiology, Elsevier Sciences, Oxford, 2002, pp. 289-309.

95.   [96] V. R. M. Correia, M. S. Carvalho, P. C. Sabroza and C. H. Vacsoncelos, "Remote Sensing as a Tool to Survey Endemic Diseases in Brazil," Cad. Saûde Pûblica, Vol. 20, No. 4, 2004, pp. 891-904.

96.   [97] B. C. Rundquist, C. J. Henrie and E. J. Grewe, "Internet Acess to Remotely Sensed Data: Satellite Imaging Made Commonplace," Journal of Map & geography Libraries, Vol. 2, No. 2, 2006, pp. 21-30.

97.   [98] S. N. V. Kalluri and T. J. Schmugge, "Application of Remote Sensing in Agriculture and Soil Science," In: K. R. Krishna, Ed., Soil Fertility and Crop Production, Chapter 18, Science Publisher, Enfield, 2002, 465 p.

98.   [99] D. Hampson, J. Crowther, I. Bateman, D. Kay, P. Posen, C. Stapleton, M. Wyer, C. Fezzi, P. Jones and J. Tzanopoulos,

"Predicting Microbial Pollution Concentrations in UK Rivers in Response to Land Use Change," Water Research, Vol. 44, No. 16, 2010, pp. 4748-4759.

99. [100] Canadian Federal-Provincial-Territorial Committee on Environmental and Occupational Health and the Canadian Council of Ministers of the Environment, "From Source to Tap: The Multi-Barrier Approach to Safe Drinking Water," 2002. http://www.hc-sc.gc.ca/ewh-semt/alt_formats/hecs-sesc/pdf/water-eau/tap-source-robinet/tap-source-robinet-eng.pdf

100. [101] Ministry for the Environment, New Zealand, "Draft Users' Guide: National Environmental Standard for Sources of Human Drinking Water," 2012 http://www.mfe.govt.nz/publications/rma/nes-draft-sources-human-drinking-water/html/index.html.

Citations

CHAPTER 1

Heejung Kim, Jin-Yong Lee, and Kang-Kun Lee, "Thermal Characteristics and Bacterial Diversity of Forest Soil in the Haean Basin of Korea," The Scientific World Journal, vol. 2014, Article ID 247401, 12 pages, 2014. doi:10.1155/2014/247401.

CHAPTER 2

Chiho Kusuda, Hikaru Iwamori, Hitomi Nakamura, Kohei Kazahaya and Noritoshi Morikawa, Arima Hot Spring Waters as a Deep-Seated Brine from Subducting Slab, doi:10.1186/1880-5981-66-119.

CHAPTER 3

Holly D Smith-Bädorf, Christopher J Chuck, Kirsty R Mokebo, Heather MacDonald,Matthew G Davidson3, and Rod J Scott, Bioprospecting the thermal waters of the Roman baths: isolation of oleaginous species and analysis of the FAME profile for biodiesel production, doi:10.1186/2191-0855-3-9.

CHAPTER 4

John E Moores, Robert H Brown, Dante S Lauretta, and Peter H Smith, Experimental and Theoretical Simulation of Sublimating Dusty Water Ice with Implications for D/H Ratios of Water Ice on Comets and Mars doi:10.1186/2191-2521-1-2.

CHAPTER 5

Bernard Thole (2013). Ground Water Contamination with Fluoride and Potential Fluoride Removal Technologies for East and Southern Africa, Perspectives in Water Pollution, Dr. Imran Ahmad Dar (Ed.), ISBN: 978-953-51-1076-7, InTech, DOI: 10.5772/54985.

CHAPTER 6

M. Rajković, I. Sredović, M. Račović and M. Stojanović, «Analysis of Quality Mineral Water of Serbia: Region Arandjelovac,» *Journal of Water Resource and Protection*, Vol. 4 No. 9, 2012, pp. 783-794. doi:10.4236/jwarp.2012.49090.

CHAPTER 7

A. Baiocchi, F. Lotti and V. Piscopo, "Conceptual Hydrogeological Model and Groundwater Resource Estimation in a Complex Hydro-thermal Area: The Case of the Viterbo Geothermal Area (Central Italy)," *Journal of Water Resource and Protection*, Vol. 4 No. 4, 2012, pp. 231-247. doi: 10.4236/jwarp.2012.44026.

CHAPTER 8

P. Turgeon, "Monitoring Recreational Waters: How to Integrate Environmental Determinants," *Journal of Environmental Protection*, Vol. 3 No. 8A, 2012, pp. 798-808. doi: 10.4236/jep.2012.328095.

Index